WHAT IS DARK MATTER?

PETER FISHER

PRINCETON UNIVERSITY PRESS
PRINCETON & OXFORD

Published by Princeton University Press
41 William Street, Princeton, New Jersey 08540
99 Banbury Road, Oxford OX2 6JX

press.princeton.edu

Library of Congress Cataloging-in-Publication Data

Names: Fisher, Peter, 1959- author.
Title: What is dark matter? / Peter Fisher.
Description: 1st. | Princeton : Princeton University Press, 2022. |
Series: Princeton frontiers in physics |
Includes bibliographical references and index.
Identifiers: LCCN 2021051624 (print) | LCCN 2021051625 (ebook) |
ISBN 9780691148342 (hardback) | ISBN 9780691185910 (ebook)
Subjects: LCSH: Dark matter (Astronomy)
Classification: LCC QB791.3 .F57 2022 (print) | LCC QB791.3 (ebook) | DDC 523.1/126–dc23/eng/20211217
LC record available at https://lccn.loc.gov/2021051624
LC ebook record available at https://lccn.loc.gov/2021051625

British Library Cataloging-in-Publication Data is available

Editorial: Ingrid Gnerlich and Whitney Rauenhorst
Production Editorial: Mark Bellis
Text and Jacket Design: Jessica Massabrook
Production: Danielle Amatucci
Publicity: Matthew Taylor and Charlotte Coyne
Copyeditor: Cyd Westmoreland

This book has been composed in Garamond & DIN 1451 LT Pro

Printed on acid-free paper. ∞

Printed in the United States of America

1 3 5 7 9 10 8 6 4 2

This book is for
Jane Ann and Olympia,
my guiding lights.

CONTENTS

INTRODUCTION: THE DARK MATTER PROBLEM 1

1. SOME BACKGROUND 5
   1.1 Mass, Weight, and Energy 6
   1.2 Distances in the Universe 12
   1.3 Measuring Speed Using Redshift 16
   1.4 Dark Energy and the Expansion of the Universe 20

2. EVIDENCE FOR DARK MATTER FROM ASTRONOMY 29
   2.1 Observations of the Coma Cluster 30
   2.2 Orbits of Stars in Galaxies 32
   2.3 Numerical Simulations of Galaxy Formation 38
   2.4 Gravitational Lensing 40
   2.5 1E 0657-56 and the Bullet Cluster 46
   2.6 Light from the Big Bang 51

3. NORMAL MATTER: THE STANDARD MODEL 63
   3.1 Particles and Interactions 63
   3.2 The Higgs Boson 68
   3.3 Testing the Standard Model 71

4. WHAT DARK MATTER IS NOT 75
4.1 Making Visible Matter: The Big Bang 76
4.2 Neutrinos as Dark Matter 86
4.3 Black Holes, White Dwarfs, Failed Stars, and Planets 88
4.3.1 Baryonic Compact Objects 88
4.3.2 Primordial Black Holes 92
4.4 Modified Newtonian Dynamics 96

5. SEARCHING FOR WIMPS ON EARTH 98
5.1 Dark Matter in Galaxies 99
5.2 Detecting WIMP Dark Matter from Elastic Scattering 101
5.3 Measuring Two Kinds of Energy 109
5.4 Detecting the Earth's Motion through the Dark Matter Halo 116

6. SEARCHING FOR DARK MATTER IN SPACE 122
6.1 WIMP Annihilation in the Galaxy 122
6.2 Detecting Cosmic Rays 127

7. SEARCHING FOR AXIONS 135
7.1 Why Do We Need Axions? 135
7.2 The Axion Dark Matter Experiment 137
7.3 The CERN Axion Solar Telescope (CAST) 141

8. EPILOGUE 146
8.1 Looking Forward: Current and Upcoming Dark Matter Experiments 146
8.2 Outlook 149

GLOSSARY 155

SUGGESTED READINGS 167

INDEX 169

WHAT IS
DARK MATTER?

INTRODUCTION: THE DARK MATTER PROBLEM

Suppose you became aware that there were specters, invisible beings, living in your house. You cannot see, hear, or feel them, but you know they are there, because they move things around your home, open and close doors, and change the room temperature. You begin to notice patterns for these changes, as if they are governed by rules.

After a time, knowing their patterns, you begin to learn the rules. You learn how to predict what changes they will make, and when they will make them. As more time passes, you come to suspect that there are many specters—maybe ten for each person in your house. The specters have always dominated your environment, and you and your family have always responded to them without knowing it.

Your curiosity about the specters grows, and you try to learn more about them—what are they made of? Where did they come from? What do they want? Still, you never sense them directly, but only learn about them through the changes they make in your (their?) home. The specters shape your environment, but you do not shape theirs. They are completely unresponsive to anything you do to communicate with or learn about them. You imagine that the specters have always been there. They are not intruders, but part of the natural order of things.

Most of us would find such a circumstance very strange, perhaps troubling, and certainly very frustrating. How could we have coexisted with so many specters for so long without knowing it? Why is it so difficult to learn about them? Where did they come from?

Over the twentieth century, **astronomers**[1] gradually became aware of "specters" in our universe in the form of a new substance first called "missing mass" and later "**dark matter**." This book uses the term dark matter. Dark matter created the shape and structure of galaxies, clusters of galaxies, and the universe itself.

The goal of this book is to make sense of the specters that represent dark matter: to explain how astronomers came to know about it; how theoreticians uncovered how dark matter shaped the largest structures in our universe through gravity; and how physicists and astronomers are navigating the complex, frustrating hunt to understand more about dark matter.

I will use the terms **visible matter** or **normal or luminous matter** to refer to **matter** that forms stars and generates the light that we observe through telescopes. Dark matter's "invisibility" means that it does not form stars or generate light (hence the term "dark" matter). More broadly, "dark" implies that dark matter does not significantly interact with visible or normal matter in any way *other* than through gravity.

Over the past 85 years, **particle physicists**, astronomers, and **astrophysicists** have shown through the process of elimination that no known substance can account for the

---

1. The glossary at the end of the book provides brief explanations of words in bold.

effects of dark matter. That includes planets, extra gas in the universe, and anything else that is made of particles that we know about. This also includes the black holes made from the collapse of stars at the end of their lives. However, there is the idea that as-yet unobserved **primordial black holes (PBHs)** that formed in the early universe from matter fluctuations in space-time could explain dark matter.

In the 1930s, a few astronomers began to understand that the amount of visible matter in clusters of galaxies could not explain the motion of the galaxies in their cluster. The total mass of the newly discovered invisible matter appeared to be tens or hundreds of times the visible mass of the stars. In the 1970s, measurements of how stars move inside galaxies led to the idea that some unseen gravitating matter causes the visible stars to orbit around the center of their galaxy faster than predicted from just the mass of the stars alone. To explain this concept, and set the stage for the rest of the book, Chapter 1 provides some physics background. Chapter 2 then lays out the evidence for dark matter from astronomical observations.

In Chapter 3, we turn to what we do know. Four **forces** describe almost all the dynamics of matter. The weak force causes radioactive decays, the strong force binds **quarks** into protons and neutrons and binds protons and neutrons into atomic nuclei, the electromagnetic force determines the structure of matter, and all matter and energy feel the force of gravity. The weak, strong, and electromagnetic forces are all variants of quantum field theory and collectively make up the **Standard Model** of particle physics. The three Standard Model forces act on quarks and leptons that make up normal matter.

The Standard Model explains almost all the observed interactions between particles made since Henri Becquerel first observed radioactive decay in 1896. Albert Einstein and his successors left us with an excellent classical theory of gravity, but theorists have been unable to find a quantum theory of gravity, leaving us with a patchwork of theories: the quantum mechanical Standard Model for quarks and leptons, and classical gravity that acts on all matter. Dark matter does not fit anywhere in our patchwork: None of the known particles from the Standard Model have the properties of dark matter; and classical gravity does not predict particles, as gravity acts on all matter.

Chapter 4 follows the experiments that led to the conclusion that dark matter does not fit into our current view of particle physics, leaving the problem of finding out what dark matter is.

Over the past 30 years, many ideas have emerged to explain the effects of dark matter. This book focuses mostly on two hypothesized new particles, called **Weakly Interacting Massive Particles (WIMPs)** and **axion**s, both of which could be dark matter particles. Chapters 5 and 6 explain some of the experiments searching for WIMPs on Earth and in space. Chapter 7 describes the idea behind axions, how axions could be dark matter, and how physicists search for axions.

This book does not end in Chapter 8 with a grand revelation of the properties of dark matter—these still elude my experimental colleagues and me. However, I hope that you will gain a deeper understanding of the dark matter problem and what a triumph it will be when we do learn something new about dark matter.

1
SOME BACKGROUND

Gravity plays a central role in everything that happens in the universe, and we will need to know a little about gravity to understand the dark matter story. The original ideas of gravity came from Galileo Galilei and Johannes Kepler in the early 1600s. Isaac Newton developed the full theory in 1687, explaining both the motion of objects acted on by a force and how massive bodies produce the forces that act on one another. In the nineteenth century, experiments began to show that Newton's laws of motion were not strictly obeyed, leading to Albert Einstein's Special Theory of Relativity in 1905, which gave universal laws of motion, with Newton's laws as approximations for objects moving at much less than the speed of light. In 1916, Einstein's General Theory of Relativity showed that gravity produced its effects through changes in the structure of space and time, with Newton's law of gravitation giving an approximation for weak gravitational forces. Einstein's theory has remained unmodified ever since.

We begin with essential background: Section 1.1 describes the relationship among mass, weight, and energy; Newton's law of gravitation; and distance scales in the universe. Mass, weight, and energy have precise meanings that are necessary for understanding how gravity works. Next, we will look at how Newton's law of gravitation exerts forces on distant objects, which will be essential for understanding how we know about dark matter. Since gravity works between distant bodies, Section 1.2 describes the typical sizes and separations of planets, stars, and galaxies in the expanding universe. Sections 1.3 and 1.4 describe the **redshift** phenomenon—crucial for measuring velocities in the cosmos—and **dark energy**.

1.1 MASS, WEIGHT, AND ENERGY

In 1687, Newton published *Principia,* in which he laid down three laws of motion. Newton's first law defines **inertia** and why objects in motion will remain in motion. The second law defines force as a change in momentum. The second law says the force on an object is the product of the object's mass and its acceleration:

$$\text{force} = \text{mass} \times \text{acceleration}. \tag{1.1}$$

We can then measure the mass of something by computing the force acting on it, measuring its acceleration, and forming the ratio of force to acceleration:

$$\text{mass} = \frac{\text{force}}{\text{acceleration}}.$$

The second law is a **kinematic law**, meaning it relates quantities of motion like velocity, force, mass, acceleration, and so on. A kinematic law tells us how objects respond to a force, while a **dynamical law** tells us what the forces are.

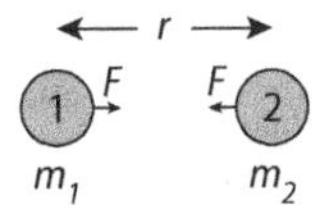

Figure 1.1. Two equal-mass bodies exert equal gravitational force on each other.

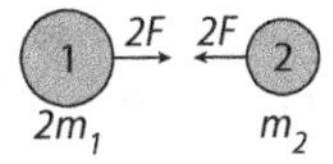

Figure 1.2. Doubling the mass of one body doubles the gravitational force felt by both bodies.

Newton's third law explains how a force acting between two massive bodies causes an equal and opposite force on each body.

Principia also set forth Newton's theory of gravity—a dynamic law of gravitational force between two bodies. While the same law applies whether the bodies are at rest or moving, for the purposes of example, if we label the bodies 1 and 2, and assume they are at rest, Newton's dynamic law of gravitation says:

- The force acting on each body has the same strength for each body and acts along the line connecting the two bodies (Fig. 1.1).
- The size of the force is proportional to the mass of each body separately—double the mass of either body, and the force acting on both bodies doubles (Fig. 1.2).
- The force diminishes as the square of the distance between the bodies—double the distance, and the force on each body goes down by a quarter (Fig. 1.3).

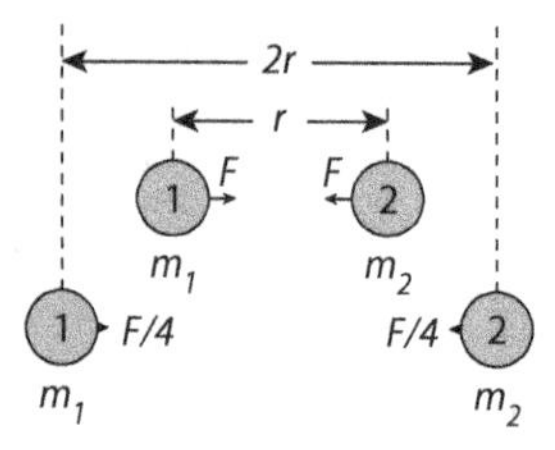

Figure 1.3. Doubling the distance between the bodies decreases by a quarter the gravitational force between them.

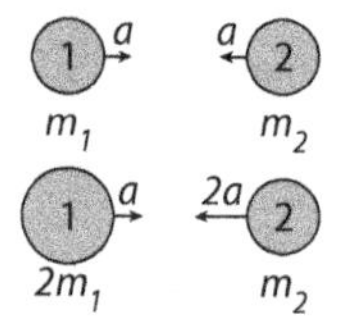

Figure 1.4. Doubling one body's mass doubles the acceleration of the other body.

When astronomers observe distant bodies moving under the force of gravity, they can measure their positions, velocities, and accelerations. We can combine the kinematic second law (Eq. 1.1) and the dynamic law shown in Fig. 1.1 to get the acceleration of the two bodies from their gravitational interaction. If we focus on body 1, the gravitational law says that the force on body 1 is proportional to its mass m_1. The second law says the acceleration, which we can measure, is equal to the force on body 1 divided by its mass m_1, which means the acceleration on body 1 from gravity does not depend on its mass. Thus if we double m_1, the mass of the first body, the acceleration of body 1 does not change, and the acceleration of body 2 doubles (Fig. 1.4).

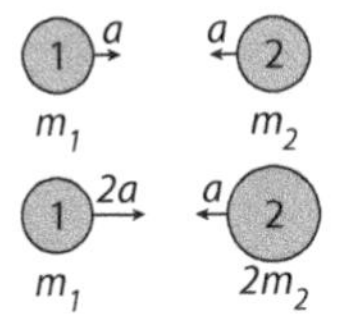

Figure 1.5. Doubling one body's mass doubles the acceleration of the other body.

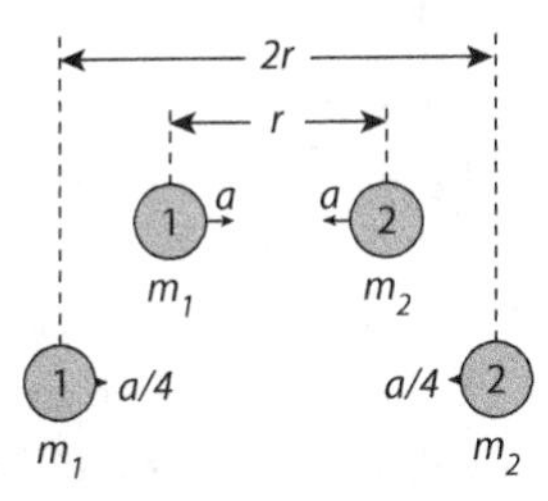

Figure 1.6. Doubling the distance between two bodies quarters the acceleration of each body.

- If the mass of the second body, m_2, doubles, the acceleration of body 1 doubles, and acceleration on body 2 remains the same (Fig. 1.5).
- If the distance between the objects doubles, the acceleration on each decreases by a quarter (Fig. 1.6).

Since the acceleration of an object from gravity does not depend on the object's mass, two objects starting at the same point in space and moving in the same direction will follow the same trajectory, even if one object is a bowling ball and the other is a feather. In the universe, this means structures like our solar system, galaxies, and clusters of galaxies stay together more than they would if they were subject to some other force for which acceleration depends on mass.[2] Think of two solar systems that are far apart in comparison with the largest planetary orbit in each, and that are stationary relative to each other. In solar system 1, the planet farthest away from solar system 2 will have a smaller acceleration than the planet closest to solar system 2 because of the diminution of its acceleration with distance. Both planets will accelerate away from the star at the center of the solar system. This is an example of a force, called a **tidal force**, acting to pull the solar system apart.

Think about using a rocket whose engine produces a force to deflect an asteroid on course to hit Earth. If the asteroid is the size of a softball, the rocket will easily push the asteroid away. If the asteroid is huge—say, the size of a ship—the rocket will push the asteroid more slowly. The mass of the ship-sized asteroid is much larger than the mass of a softball-sized asteroid; so, given the same force, the smaller asteroid will move out of the way faster.

People sometimes say "weight" when they mean "mass." Weight expresses the force of gravity on an object near the Earth's surface. We measure weight in pounds and convert pounds to kilograms by dividing the weight by 2.2 pounds per kilogram, which is fine on Earth. If we measured the weight of *The Physics of Energy* by Jaffe and Taylor (an excellent book, and the second-heaviest book I own) using a spring scale, we would find it weighs 5 lb. 13 oz., or 5.8 lb. A spring scale measures the amount of gravitational force acting on a body by balancing the gravitational force against

2. For example, the electron and proton in a hydrogen atom are held together by electromagnetic force. The average distance between the electron and proton is 200 times greater than it would be if the electron was replaced by its heavy cousin, the muon.

the force a spring must apply to keep the object from falling. The book's weight, 5.8 lb., is a measure of the force that the Earth's mass exerts on the book and is equal to the book's mass times the gravitational acceleration near the Earth's surface, 9.8 m/s^2. On Earth, the conversion factor from the force measurement in pounds to a mass in kilograms works out to 0.455 kilograms/pound, so the book has mass of 2.6 kg.

If we took the book to the Moon, its mass would still be 2.6 kg. However, near its surface, the Moon has a gravitational acceleration of 1.63 m/s^2, or 1/6 of Earth's. If we used the same spring scale as we used on Earth, it would measure 1/6 as much force on the Moon as on Earth, and the book would appear to weigh just under 1 lb. Weight depends on the force of gravity at the object's location, while mass remains the same everywhere.

If you have some knowledge of special relativity, you will notice the expressions above are classical, not relativistic. I mostly use the classical expressions here, because they are accurate enough just about anywhere we will need them in our exploration of dark matter.

A rich array of particles and radiation fills the universe. When we talk about the amount of "stuff" in the universe, we can talk about mass and energy interchangeably: Einstein's famous equation, $E = mc^2$ (which states that an object's energy is equal to its mass times the square of the speed of light), tells us that the energy of a particle at rest is proportional to its mass. Not all particles in the universe are at rest, but most move so slowly that the energy associated with their motion is very small compared to the energy associated with their mass, and $E = mc^2$ is a perfectly good approximation. This book uses **mass-energy** or just "mass" to signify the stuff of the universe, which will include the mass-energy of the matter in planets, stars, and galaxies.

1.2 DISTANCES IN THE UNIVERSE

We will talk much more about objects in the universe throughout the book, but this is a good place to get a handle on the sizes of things.

First, a grounding in the units of measure we use in calculating distances in the universe. Astronomers, astrophysicists, and **cosmologists** observe and study objects in the universe over a large range of distances. The nearest natural extraterrestrial body is the Moon at 239,228 miles, or 385,000 km, and the longest is to the "edge of the universe," the distance traveled by light just reaching us now from the start of the universe 13.8 billion years ago. The ratio of the longest to the shortest distances scientists studying the universe observe is 300,000,000,000,000,000, or 3×10^{17}. One could use meters and **scientific notation** to express any distance, but different subfields have developed different units of distance still in use today. The two most common are the **astronomical unit (AU)** and the **light-year** (lt-yr). The astronomical unit is the distance from the Earth to the Sun, and 1 AU = 149,500,000 km = 92,956,000 miles. For the light-year, light travels at 300,000 km per second, and a year is 31,536,000 seconds, so a light-year is 946,000,000,000 km = 5,879,000,000,000 miles. It is not critical to remember these calculations, but it is helpful to understand the basis for these units of measure.

Next, let us start the discussion of size with the Earth, whose average radius is 6,371 kilometers. The Earth orbits the Sun in an almost circular orbit at a radius of 1 AU. The most distant planet, Neptune, orbits at an average distance of 30 AU, and the dwarf planet Pluto lies at 40 AU on average (Pluto's orbit is not circular, and sometimes Pluto is closer to the Sun than Neptune is). The Earth-Sun orbit is the right unit for measuring things in planetary systems whose sizes are less than 100 AU. The Sun's nearest neighboring star is Proxima Centauri, which lies 267,100 AU or 4.2 lt-yr from Earth. For interstellar distances, a larger unit of measure is helpful, and we will use the light-year.

Gravitationally bound groups of stars form the visible parts of galaxies. Our sun and Proxima Centauri are just two of about 100 billion[3] stars in the Milky Way galaxy (Fig. 1.7). The stars in the Milky Way inhabit a disk about 100,000 lt-yr, or 100 klt-yr, in diameter and less than 3 klt-yr thick. The Sun lies about 25 klt-yr from the center of the Milky Way. A **spiral galaxy**, the Milky Way itself lies about 2.5 million light-years—2.5 Mlt-yr—from Andromeda, another spiral galaxy. The Milky Way also has two small galaxies close to it, the **Large and Small Magellanic Clouds (LMC and SMC)**, about 160 klt-yr and 206 klt-yr away, respectively.

A group of galaxies may be gravitationally bound, forming a galaxy cluster. Galaxy clusters typically have 50 to 1,000 galaxies spread across up to 5 Mlt-yr. "Superclusters,"

3. In the United States, "billion" refers to a thousand million; while in the UK, billion refers to a million million. In this book, we will use billion in the U.S. sense but will use the standard metric prefix G (for "Giga") for a billion. A billion light-years will be written as 1 Glt-yr.

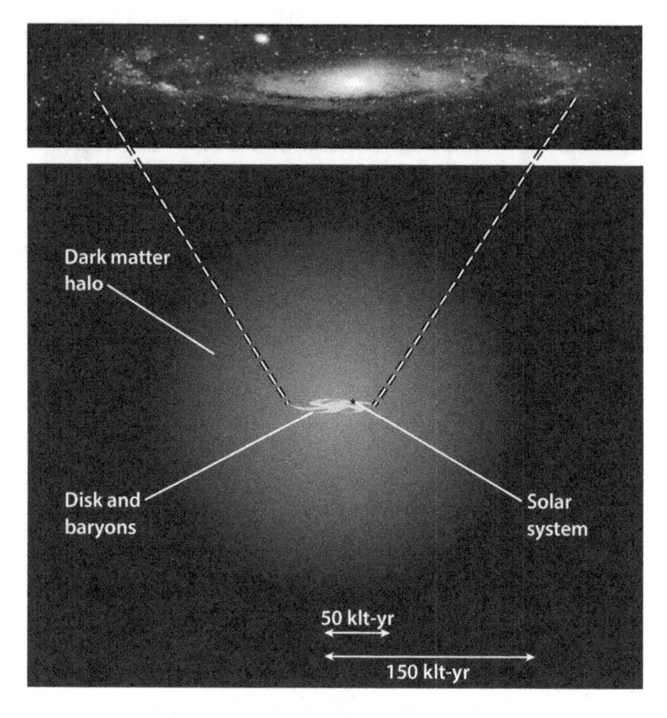

Figure 1.7. Galactic structure. The upper panel is an image of Andromeda, the nearest galaxy to the Milky Way. The lower image shows a model of a galaxy that includes a spherical region of dark matter extending well beyond the visible disk of the galaxy. The spherical dark matter halo explains the motion of the stars in the galaxy, as we will see in Chapter 2. The star indicates the approximate position of our solar system in the Milky Way, about halfway across the visible disk.

gravitationally bound clusters of galaxy clusters, are up to 500 Mlt-yr across and are the largest structures in the universe.

Figure 1.8 shows these and other distances all together. At every distance scale, from the solar system to galactic

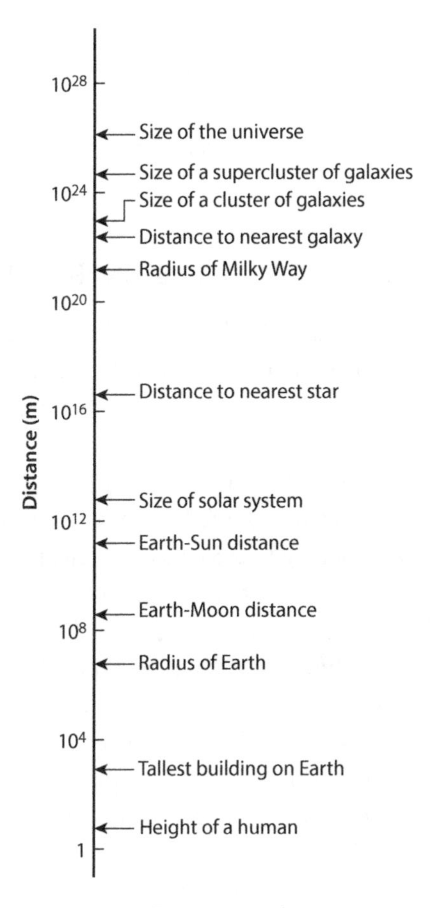

Figure 1.8. Distances in the universe.

superclusters, the gravitational force holds the components in a closed orbit around each other. Thus for any gravitationally bound object—for example, a star in a galaxy—the object's kinetic energy is smaller than its gravitational potential energy, so the object cannot escape to an infinite distance and instead follows an orbit. The total mass of a system determines the gravitational binding energy, so the mass of a system determines the size of the system. In Chapter 2, we will look at how dark matter determines the size and shape of the different structures in the universe.

1.3 MEASURING SPEED USING REDSHIFT

Electromagnetic waves (or light) travel at the same speed, 186,000 miles per second, regardless of the motion of the emitter or observer. Quantum mechanics treats light as both a wave and a particle, called a photon.

If we arrange to collect light for long enough that many photons arrive and are counted, we can think of light as a wave. For example, the Hubble Space Telescope (HST) collects about 100 million photons per second when pointed at a Milky Way–sized galaxy over 60 Mlt-yr away, so it makes sense to think of HST as working with waves for this observation.

A wave has crests and troughs, where the electric and magnetic fields that make up the wave point in one direction or the other. The wave moves at the velocity of light, so if you label a crest from a stationary emitter and follow its motion, you will see it move with the speed of light along with the wave (upper and lower curves in Fig. 1.9A).

Suppose the emitter moves away from the viewer. The distance between successive crests will be longer than if the emitter were stationary, because the emitter moves between the emission of the two crests (upper and lower curves in Fig. 1.9B, observer on left side of figure). The whole wave gets stretched out this way, and the amount

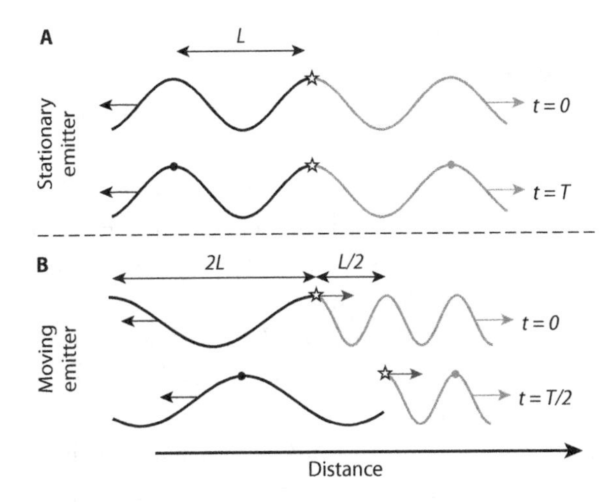

Figure 1.9. Light waves from a stationary and a moving emitter. Panel A shows the emission of light waves from a stationary emitter, indicated by the star. The lower curve shows the waves at a time one period T later. The crests, indicated by the dots on the lower curve, move to the left and right at the same speed, and both waves have wavelength L. Panel B shows the emission of light waves from an emitter moving to the right with 3/5 the velocity of light. The lower curve shows the waves a time $T/2$ later, where T is the period of the waves in Panel A. The waves traveling to the left are stretched out to a wavelength of $2L$, and those traveling to the right are compressed to $L/2$. The crests still travel at the speed of light. The right-going wave has twice the frequency of the waves in Panel A, and the left-going wave has 2/3 the frequency of the waves in Panel A.

of stretching is proportional to the velocity with which the emitter moves away from the viewer.

The distance between the crests is called the wavelength of light, L, and determines its color, with longer wavelengths being red and shorter wavelengths being blue or violet. An emitter moving away from the receiver will stretch the emitted light, making the wavelength longer and moving the light closer to the red end of the spectrum. This is why the effect is called the redshift.

If the emitter moves toward the viewer, the wave is compressed and shifted toward the blue end of the spectrum (upper and lower curves in Fig. 1.9B, observer on right side of figure). This is called the blueshift.

Einstein's Special Theory of Relativity shows that a moving clock appears to tick more quickly, which also causes a shift in the wavelength. This is why there are two crests on the right side of Figure 1.9B, bottom curve, while there is only one on the left. The situation is more complicated if the emitter moves at an angle to the viewer, but the result is the same: An object moving away means an object appears to be redder, and an object moving toward the viewer becomes bluer.

Astronomers can use the redshift to determine the velocity of a distant object. This is possible because most stars that are viewed with telescopes emit light with a specific wavelength (Fig. 1.10). For example, hydrogen emits some light with a wavelength of 91 billionths of a meter. If a galaxy is moving away from Earth at 1% of the speed of light, the emitted wavelength will be stretched out by 1% toward the red end of the spectrum, and this can be measured with a telescope attached to a spectrometer. By measuring this wavelength and comparing it with the wavelength of hydrogen in a stationary laboratory, the velocity with which the galaxy moves away from the Earth may be determined. Vesto Slipher (Lowell Observatory) first used redshifts to measure motion of stars inside galaxies.

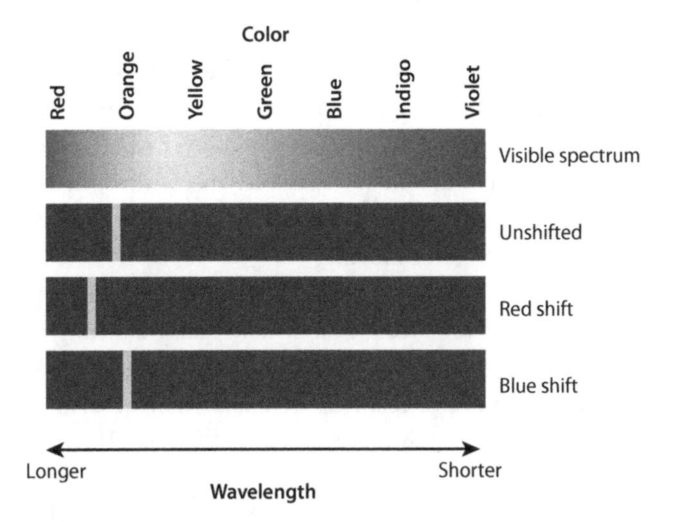

Figure 1.10. The spectrum of visible light and the effects of red- and blueshift. Top panel shows the visible light spectrum from red (longer wavelength) to blue (shorter wavelength). The second panel shows the light at a specific wavelength, called a transition line, emitted in an atomic transition in a star. The lower two panels show the red- and blueshifts of the transition line, respectively.

Redshift is the technique that astronomer Edwin Hubble used in determining the expansion of the universe and later was used in determining the velocity of rotation of stars within nearby galaxies. Redshift is a commonly used technique, and the attainable precision can be very high—some velocities of planets orbiting nearby stars are measured with accuracies of 1 m/s.

The phenomenon of redshift is similar to the Doppler shift of sound that we are well familiar with—for example, the sound of the siren of an approaching or receding police car shifting to higher or lower frequency, respectively. There is an important difference, though, for a sound wave: The velocity of sound is always relative to the medium carrying the sound—the air. A stationary observer will measure a different velocity for sound than an observer who is riding along in the police car, for example. In contrast, light is observed to move at the same velocity for every observer, whether stationary or moving.

The stationary observer will measure the same velocity as the observer sitting in a distant galaxy emitting the light that the stationary observer on Earth sees. This is an important feature of relativity and was a key observation that Einstein made in constructing his Special Theory of Relativity.

1.4 DARK ENERGY AND THE EXPANSION OF THE UNIVERSE

The story of dark matter takes place in an expanding universe. Right now, the dominant component of the matter and energy in the universe is another dark substance called dark energy. Dark matter and dark energy may be in some way related, but no one yet knows how, and the effects of the two forms of matter-energy are very different. Dark matter shapes galaxies and clusters of galaxies, and it helps dictate how the universe expands. Dark energy affects the expansion rate of the universe; and the current data imply that dark energy's effects are the same everywhere in the universe and that dark energy acts as a **cosmological constant**, meaning dark energy's pressure is the opposite of its density. This section explains how dark energy drives the expansion of the universe, assuming that dark energy acts as a cosmological constant.

What does it mean for the universe to "expand"? The term implies that the universe gets larger somehow, but what does that mean for something infinitely large?

Exploring how the universe expands requires measuring both the distance and velocities of distant **nebulas**[4] or galaxies. Measuring the reddening or redshift of the light from a galaxy provides the velocity of the galaxy moving away from Earth. Distance measurement to galaxies requires "standard candles"—astrophysical objects in the distant galaxy with predictable light output, such as **Cepheid variable stars** first noticed by Leavitt.

In the early 1920s, astronomers were divided on the question of the nature of nebulas—fuzzy-looking astrophysical objects. Some thought nebulas were clouds of gas inside the Milky Way, while others believed nebulas were collections of stars (like our Milky Way) that lay far outside it.

In 1924, Hubble measured the distance to the nebula Andromeda using Cepheid variable stars and, combined with redshift measurements made by American astronomer Vesto Slipher, showed that galaxies are collections of stars millions of light-years from—and not gas clouds in—the Milky Way. In 1929, Hubble measured the redshifts of about 24 galaxies and found that most were moving away from Earth. Hubble used Cepheid variable stars and, in some cases, assumed that the galaxies were all of the same luminosity to estimate their distance from Earth. He found a proportionality between how fast they moved away from

4. Until the early twentieth century, galaxies were referred to as "nebula," which means "foggy" in German.

Earth and their distance from Earth. The proportionality became known as Hubble's Law, and the proportionality constant known as the Hubble Constant, H_0.[5]

Hubble used the redshift to measure the velocities of the galaxies he observed. The relationship between the redshift and velocity was well known from Einstein's Special Theory of Relativity, but Hubble's galaxy recession measurement is caused by the expansion of the universe that relates distance from a galaxy to an observer on Earth and the redshift. Einstein's Theory of General Relativity relates distance to the redshift through the notion that the universe is expanding or contracting in a way governed by the matter-energy content of the universe. In Einstein's theory, the observed galaxy is not moving *through* space; instead, space is expanding, carrying the galaxy with it.

This still sounds strange. Think about a half-inflated balloon with a grid marked on it, and zoom in to a very small patch of the surface of the balloon with an ant on it. If the size of the patch is small enough (say, 1% of the radius of the balloon), then the patch will seem flat to the ant. For all practical purposes, the ant will think it is on a flat patch of balloon, just the way we don't notice the Earth is a sphere when we drive from our house to the grocery store (Fig. 1.11).

If the ant walks on the surface, we can count how many grid squares it crosses in a second and find the velocity of the ant relative to the grid. Now think about two stationary ants (relative to the grid) and how they "move" relative to each other when the balloon doubles in size. In terms of

---

5. Once the Big Bang was better understood, it became clear that the Hubble Constant varied with time. H_0 and the term Hubble Constant refer to its value today.

motion relative to the grid, the two ants remain stationary. However, if the second ant emits light, observed at the location of the first ant, the first ant will measure the light as redshifted and could reasonably conclude that the second ant was moving away. This is a different kind of motion than the motion of the ant relative to the grid. The expansion of the surface of the balloon causes the motion inferred by the redshift measurement of the first ant.

On an expanding spherical balloon, every observer measures the same thing from the surrounding objects: Objects move away from the observer's location as the balloon expands. The relative motion of the two objects does not depend on direction on the balloon.

The universe expands similarly: Cosmologists assume that the relative motion does not depend on position in the universe or the direction between two objects. Einstein showed that matter and energy are equivalent, and he applied this notion to show that the expansion rate of the universe depends on the total matter and **energy density** in the universe, which he assumed to be the same everywhere. The apparent motion between two objects caused by the expansion of the universe, called the recession velocity (v_{rec}), divided by the distance between the two objects, is proportional to the **matter density**, or

$$v_{rec} = \text{constant} \times \sqrt{\text{density}} \times \text{distance}.$$

The total matter density and energy density in the above equation is denoted ρ and is the sum of the densities of all the different kinds of matter and energy:

$$\rho = \rho_{\text{DarkEnergy}} + \rho_{\text{DarkMatter}} + \rho_{\text{NormalMatter}} + \rho_{\text{Light}} + \dots.$$

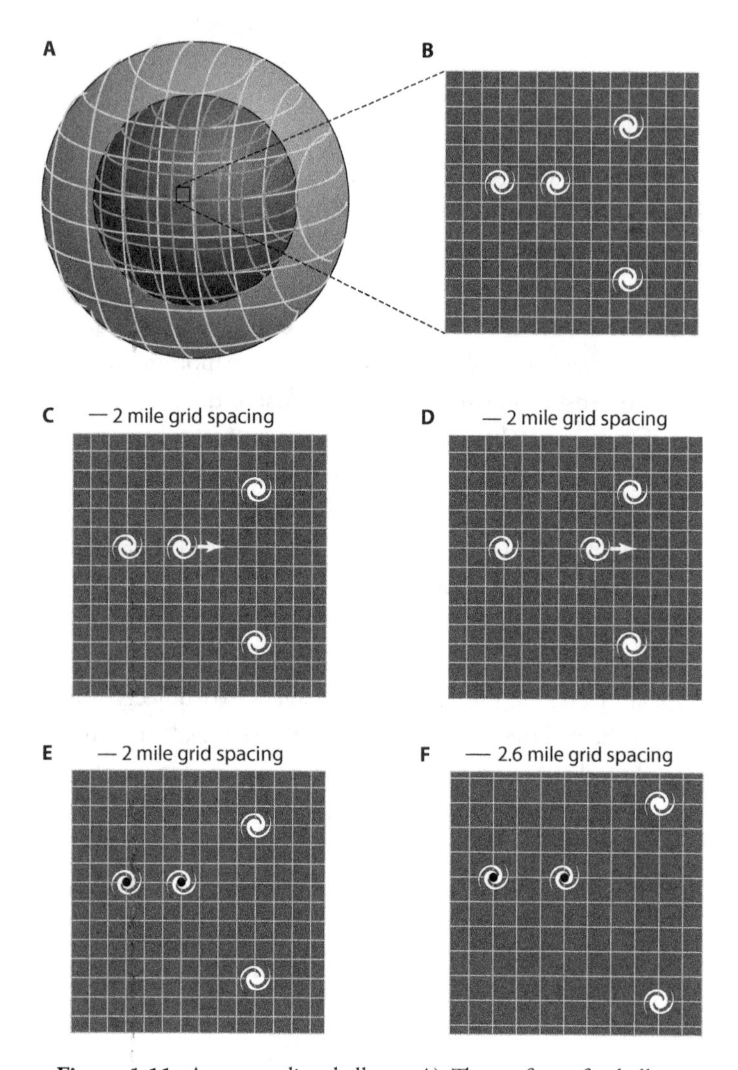

Figure 1.11. An expanding balloon. A) The surface of a balloon with grid lines marked at two different times: one at the start of the scenario, and a second when the diameter of the balloon has doubled since the first time. The grid lines are curved because the surface is not flat. B) A very small patch of the surface—small enough so it seems flat, like the surface of the Earth to us when we stand on it. The spirals indicate objects scattered on the surface.

The densities of different types of matter and energy change as the universe expands. For visible and dark matter, there are a fixed number of particles initially, so the number density—number of particles in a volume—decreases as the universe's volume increases with the expansion. The number density of light particles—photons—also dilutes as the universe expands, but the light waves also stretch out with the expansion. Stretching the light waves lowers their wave energy, and the decrease in light matter-energy density is greater for a given amount of expansion than the decrease in the density for visible and dark matter. Light waves carry energy stored in the bending of the electric and magnetic fields that make up the light—the sharper the bends in the fields, the more energy the fields store. You can see this by crumpling a piece of paper loosely into a ball. The crumpling stores energy from your hands squeezing the paper into creases in the paper. Crumpling a second piece of paper more tightly into a ball smaller than the first will result in more sharper creases, which you will see when

Figure 1.11 (continued). C) An object moving on the surface, indicated by the arrow. Say that each grid line is 2 miles from the next. D) The same patch, an hour later. We can find out how fast the object is moving by counting how many grid lines are passed (2 grid lines), dividing by how long it took to move that far (1 hour) to get the speed: 4 miles per hour. E) The same small patch of the balloon with the objects restored to their original locations. Now the balloon is expanding. Two objects are marked with a black dot. F) When, an hour later, the balloon has expanded by 30%, each object is 30% farther apart. In the case of the two marked objects that were 3 grid lines (or 6 miles) apart and are now 7.8 miles apart, the change in distance is 1.8 miles, so the objects move apart at 1.8 miles per hour due to the expansion of the balloon.

you unfold both pieces of paper and compare them. The expansion of the universe stretches out the creases in the electric and magnetic fields over a larger distance, smoothing them out and reducing the amount of energy the fields store.

In contrast, the dark energy density remains constant as the universe expands. Unlike a fixed number of particles in a growing box, the density of dark energy does not change as its box expands. With this in mind, think about a universe with only dark energy in it. The recession velocity v_{rec} is how much the distance, d, between us and a distant galaxy changes each second; and v_{rec}/d is the fractional recession velocity—by what fraction the distance changes each second.

The fractional recession velocity is analogous to the interest rate on a bank account: If you put \$100 into a bank account earning 7% interest per year, after 1 year, you will have a bit more than \$107 in your account. If a galaxy is 1 Mlt-yr from Earth and the fractional recession velocity is 7%/Glt-yr, then after a billion years, the galaxy will be a bit more than 1.07 Mlt-yr away. The "a bit more" comes from the compounding of the "interest." A feature of both compound interest and expansion of the universe driven by dark energy is that in both cases, the "principal" (amount of money in the account or distance between galaxies) grows without bound.

Right now, the best measurements that astronomers have made imply that dark and normal matter comprise 31% of the energy of the universe, and the other 69% is dark energy. The presence of both dark and normal matter and dark energy makes the time evolution of the universe more complicated than in a universe with only dark energy. As the universe continues to expand, the matter density will decrease, while the dark energy density will remain the same. This means the matter fraction will decrease as the matter dilutes, causing the dark energy fraction to increase. Eventually, the matter fraction will become insignificant compared to the dark energy fraction, and the universe will expand exponentially.

Dark energy affects the universe as a whole. The next-smallest structure, the galaxy cluster, is a group of galaxies held together by the gravity pulling each galaxy toward all the other galaxies, which is much stronger than the expanding universe trying to pull the galaxies away from one another. The cluster as a whole follows the expansion driven by dark energy, called the **Hubble flow**. The galaxy cluster does not change in size and moves along with the Hubble flow as viewed by an observer on Earth. Imagine river rapids that move ever faster along the river's course. A raft moving down the rapids would feel a stronger tug from the faster moving water on the front trying to stretch the raft, compared to the slower moving water at the back; but the material of the raft resists the stretching force, so the raft does not change length as it moves along the river.

Einstein's early application of this theory of gravity to the universe included something like a cosmological constant. Einstein added this concept to his theory when he found that matter in the universe would cause it to expand, while the cosmological constant would keep the universe from expanding. This was consistent with the thinking at the time, as most astronomers believed the universe was static. Subsequently, over the course of 10 years, Hubble's measurement showed that the universe was expanding, removing the need for the cosmological constant.

By the late 1980s, the version of cosmology and the Big Bang theory—a model of the formation of the matter in the universe—included Hubble's Law, which says that the velocity with which a galaxy appears to move away from us is proportional to its distance from us. Measurements of distant supernovas and the light from the Big Bang made in the 1990s showed that galaxies halfway across the visible universe receded faster than predicted by Hubble's Law. The relationship is consistent with the 7% interest example described earlier in this section, and dark energy causes the faster expansion. Later measurements showed that dark energy constitutes about 69% of the matter and energy content of the universe.

The specters mentioned in the Introduction were the dark matter, invisibly shaping our galaxies and larger structures of the universe. In this analogy, dark energy shapes the universe as a great, slow explosion, gradually pushing everything apart, the same way everywhere. The next chapters explain how dark matter shapes galaxies and clusters of galaxies, and how dark energy causes a slow, majestic, inexorable expansion of everything.

Dark energy poses a problem for **particle theory**: No known particle can be the dark energy particle, and there is no obvious extension to our current theory that can explain dark energy. But Einstein's General Theory of Relativity can accommodate dark energy as a version of the cosmological constant. Chapter 4 will delve into this theoretical situation more fully.

2
EVIDENCE FOR DARK MATTER FROM ASTRONOMY

This chapter explains why we think dark matter exists and how dark matter influences the shapes of galaxies, clusters of galaxies, and the universe. All the evidence for dark matter comes from astronomical observations of distant objects the size of a galaxy or larger.[6] What we know about dark matter only comes from measuring how the gravitational force exerted by dark matter on the visible matter in these large, distant objects changes how big they are, how they are shaped, how fast they rotate, and how fast they move relative to one another—much like the specters in the Introduction, moving things around the house.

There are three other forces in the universe: electromagnetic, weak, and strong forces. As far as we know, dark matter is not significantly affected by those forces, which will be explained more in Chapter 3. Dark matter might, for example, feel the electric force, but at a level one 10 billionths weaker than the interaction between an electron and a proton in a hydrogen atom. Everything we know

---

6. A typical galaxy is about 150 klt-yr across, contains about 10–100 billion stars, and lies about 2 Mlt-yr from the nearest neighboring galaxy.

about dark matter comes from studying the gravitational interaction between dark matter and known matter.

2.1 OBSERVATIONS OF THE COMA CLUSTER

In 1933, the notion of a galaxy as a collection of stars outside the Milky Way was less than 10 years old when astronomer Fritz Zwicky at Caltech wrote a paper describing measurements of the total mass of the **Coma Cluster**—a nearby (336 Mlt-yr from us) gravitationally bound assembly of about 1,000 galaxies. Measuring the mass of an object composed of 100 trillion or more stars moving under their mutual gravity presented a challenge to observers.

Zwicky used the motion of galaxies in the Coma Cluster to estimate the gravitational mass of the cluster, and he measured the light output of all the galaxies in the cluster. He compared the mass-to-light ratio for the Coma Cluster with that for Kaptyn's star, choosing Kaptyn's star because it lies outside the Milky Way and is relatively isolated. When Zwicky compared the mass-to-light ratios, he found that the stars in the Coma Cluster emitted much less light per star than Kaptyn's star.

Zwicky made his survey of the thousand or so galaxies of the Coma Cluster using the Schmidt-Cassegrain refracting telescope at the Palomar Observatory in California. Although only 18 inches in diameter, the Schmidt-Cassegrain telescope had a large (2°) field of view, making it ideal for surveys. The Coma Cluster covers about 4° × 4° on the sky (for comparison, the Sun and Moon both are about 0.5° across) and lies about 336 Mlt-yr from the Earth. On average, the galaxies are about 1 Mlt-yr apart,

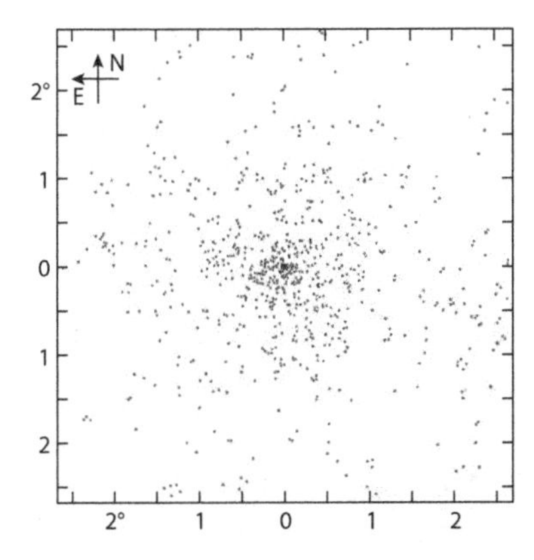

Figure 2.1. Sky map of the Coma Cluster. Inverted image of the Coma Cluster from Fritz Zwicky's paper. The image shows the brightest 200 of the more than 800 galaxies in the cluster.
Source: *Astrophysical Journal*, l. 86, No. 3, p. 217, 1937.

although they are closer to one another at the center of the cluster. Zwicky's survey gave the position of each galaxy in the sky (Fig. 2.1).

First, Zwicky measured how the galaxies were distributed around the center of the cluster, as shown in Figure 2.1. Next, he measured the redshifts of about ten galaxies in the Coma Cluster and found that they move with a spread of velocities, or **velocity dispersion**, of about 700 kilometers per second.

Zwicky analyzed the Coma Cluster as if it were a gas, with the galaxies playing the role of atoms. This approach works because the galaxies in the cluster do not collide with one another much, and their aggregate gravitational attraction determines their motion. Under these circumstances, a physics theorem, the **virial theorem**, relates the average kinetic energy of the cluster (which depends on the average and dispersion of the velocities of the galaxies) with the cluster's potential energy (which depends on the total mass of the cluster).

Having measured the average and spread in velocities, the virial theorem then gives the mass of the cluster, so Zwicky could form a ratio of mass-to-light output for the cluster. The ratio came to about 400–500 solar masses per solar luminosity.

Observation of Kaptyn's star gave a mass-to-light ratio of about 3 solar masses per solar luminosity, leading Zwicky to conclude that a substantial fraction of the matter in galaxies is not light-emitting stars in the galaxies but something else.[7] Zwicky's observation was the first evidence that a substantial fraction of the mass of a galaxy did not emit light. The nature of this dark matter was not at all clear at the time.

In the years after Zwicky's discovery, there were various speculations for the unaccounted-for mass: For example, gas or dust between the galaxies, failed stars that do not emit light, or stars between the galaxies. The main conclusion was that galaxies had a substantial fraction of their mass tied up in something other than light-emitting stars.

2.2 ORBITS OF STARS IN GALAXIES

Following Zwicky's work, more studies of the motions of galaxies in clusters implied that most of the mass in other

---

7. In his 1933 paper, written in German, Zwicky wrote, "the surprising result would emerge that dark matter is present in very much greater density than luminous matter." In a more famous 1937 paper in the *Astrophysical Journal*, he is much more circumspect, "This discrepancy is so great that further analysis of the problem is in order." The term "dark matter" does not appear in the 1937 paper.

Figure 2.2. The galaxy Andromeda. Andromeda, also known as M31, is the Milky Way's closest galactic neighbor, at 2.5 Mlt-yr away. The image shows an area about 75 klt-yr across.

galaxies was not in light-emitting stars, supporting Zwicky's observations. The motion of stars in galaxies offered another means of measuring a galaxy's mass. The motion of the constituents of a "spiral nebula" was first observed in 1914 by Slipher, and measurements have been made with increasing accuracy ever since. The early work centered on the Andromeda and Triangulum galaxies, because they are closest to Earth and the easiest to observe.

Andromeda, the closest large galaxy, lies about 2.5 Mlt-yr away from Earth. As a spiral galaxy, Andromeda has a bright central region and more diffuse spiral arms extending outward 70 klt-yr from its center (Fig. 2.2). Farther out, hydrogen gas has been detected from radio measurements.[8] From Earth, we view Andromeda at an angle of 13° from edge-on, making the structure of the galaxy apparent. Along with about 80 other galaxies, dwarf galaxies, and the Milky Way, Andromeda forms the Local Group of galaxies that extend across about 10 Mlt-yr. The smaller galaxy Triangulum, second closest to Earth, also is a member of the Local Group.

For a star in a circular orbit around the center of a galaxy, the amount of mass inside the star's orbit determines its orbital velocity. Newton's laws imply:

1. The orbit velocity depends on the square root of the total mass (normal plus dark) inside the orbit radius, and
2. The orbit velocity depends on the reciprocal of the orbit radius.

Astronomers want to find the total mass of the galaxy and how that mass depends on the square root of distance from the center of the galaxy. One way to do this is by using redshifts to measure the orbital velocities at a succession of closely spaced radii. The difference in velocity between two points close in radius depends on the distance between the two measurements, the total mass inside the smaller radius, and the amount of mass in between the radii of the two measurements. If measurements are made near the center of the galaxy where the radius is small, we can assume that the mass inside that radius is given by the number of stars, which we can measure. Then we can work our way out,

8. Hydrogen emits a faint radio signal related to its temperature. The redshift of the radio waves from the hydrogen gives the velocity of the hydrogen.

point by point, and build up a mass profile of the galaxy from each measurement of the mass between the successive orbit velocity measurements. This process is tricky: Galaxies have complicated shapes, requiring one to make certain assumptions about the distribution of the masses and their velocities inside the galaxy.

Before the 1970s, measuring the velocity of gas or stars in a remote galaxy was quite difficult. A typical star in Andromeda or the Milky Way moves at approximately 200 km/s, making the precise measurement of a 0.1% change in the wavelength of the emitted light challenging. A large telescope collects light of all colors roughly equally and focuses the light to create an image of a distant object on the image plane of the telescope. Measuring the velocity of a particular region of the image requires that the light from that region pass through a slit in the image plane of the telescope and be projected on a spectrometer, which spatially separates the different wavelengths of light through a prism or diffraction grating and allows measurement of the light's intensity as a function of wavelength. The difficulty arises because there is not much light going through the slit, and long exposure times can be necessary to collect enough light on photographic film. For example, in 1937–1938, Horace Babcock (Berkeley) used exposure times as long as 21 hours in his study of the rotation of Andromeda. As that spanned several nights of observation, there was the added complexity that the telescope had to be aimed and aligned precisely the same way on successive nights, which required even more observing time.[9]

---

9. A particular telescope is likely to have 200 clear observing nights per year. Taking into account full moons, that allows for 150 nights. Each night is about

Measurements of the velocities of stars in galaxies as a function of radius were first carried out starting in the late 1930s by Babcock, Walter Baade (Mt. Wilson Observatory), and Nicholas Mayall (Lick Observatory). Their measurements indicated that stellar velocities may be higher than expected from the amount of light emitted by the stars in the host galaxy. In 1959, Louise Volders measured the velocity of hydrogen gas in the nearby Triangulum Galaxy using the redshift of the light emitted by the atomic transitions of hydrogen. From Newton's second law, the velocity of the gas orbiting around the center of Triangulum should be proportional to the amount of matter inside the orbit of the gas. Volders found the hydrogen gas has a much higher velocity than expected from the number of stars inside the orbit in Triangulum, indicating there must be more mass than just the mass of the stars.

Working in 1969, Vera Rubin and Kent Ford (both at Carnegie Observatories) enjoyed the advantage of a new image intensifier tube that allowed good redshift data to be taken in 60 to 90 minutes, greatly increasing the number of measurements possible. They chose to use large regions of singly ionized hydrogen, called HII regions, that emit light as the atomic electron de-excites in hydrogen. Their observations ranged from close to the center of the galaxy to the edge of the luminous region at 15 klt-yr. Mort Roberts (Green Bank Observatory) used radio emission from the hydrogen to extend the velocity measurements outside the luminous region to 50 klt-yr. The typical

---

6 hours or fewer. Twenty-five hours of observing time is thus 3% of a telescope's time in 1 year.

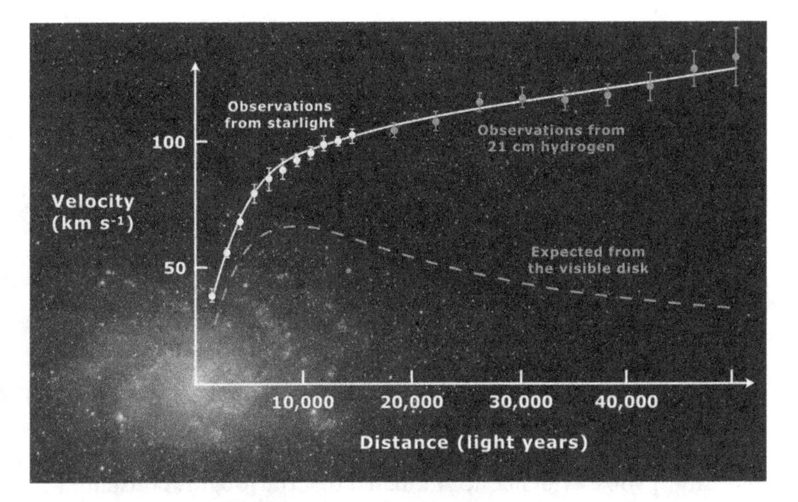

Figure 2.3. Rotation curve of Triangulum (M33). The *x*-axis is the distance from the center of the galaxy, and the *y*-axis is the radial velocity as determined from the redshift measurements of the hydrogen gas in the galaxy. The extent of the visible light is indicated by the image of the galaxy lying along the *x*-axis. The solid line indicates the measured velocities as a function of radial distance, and the dotted line indicates the expected velocity if the stars were the only massive objects in the galaxy. The rotation curve does not vary much with radius past 20 klt-yr and is referred to as "flat." It extends well past the end of the visible galaxy. This indicates that dark matter extends at least to 50 klt-yr in radius from the center of the galaxy (Public domain).

velocities of the observed regions were 100 to 250 km/s (Fig. 2.3).

The combined radio and optical data showed that the orbital velocity did not depend much on radius for radii from around the bright central core at about 10 klt-yr to

150 klt-yr. The measurements showed that there must be substantial mass well beyond the radius of the visible stars. Computing the total mass of Andromeda, the rotation curve measurements of Rubin, Ford, Roberts, and others found a mass for Andromeda of about 1.5 trillion times that of the Sun, with the total light output being about 34 billion times that of the Sun. As with the earlier measurements of the Coma Cluster, over 90% of the mass of Andromeda was not associated with light production. They also found that the dark matter components in Andromeda extended well beyond the visible region of the galaxy.

As more measurements were made during the 1970s and 1980s, evidence accumulated that the roughly constant orbit velocity outside the center of galaxies was a typical feature of galaxies and was not unique to Andromeda or Triangulum, as shown in Figure 2.4.

2.3 NUMERICAL SIMULATIONS OF GALAXY FORMATION

Measurements of the rotation of many more galaxies showed that their orbital velocity was nearly independent of the radius well outside the luminous disk, exhibiting the same approximately 90% mass deficit compared to the mass expected from the light measured from stars in Andromeda. In the mid-1970s, new minicomputers gave astronomers the ability to simulate the evolution of galaxies by solving Newton's Law numerically for hundreds or thousands of mass points. The simulations showed that, without the inclusion of about 10 times as much dark matter as normal matter in a spherical halo extending well outside the region of normal matter, the mass points would very

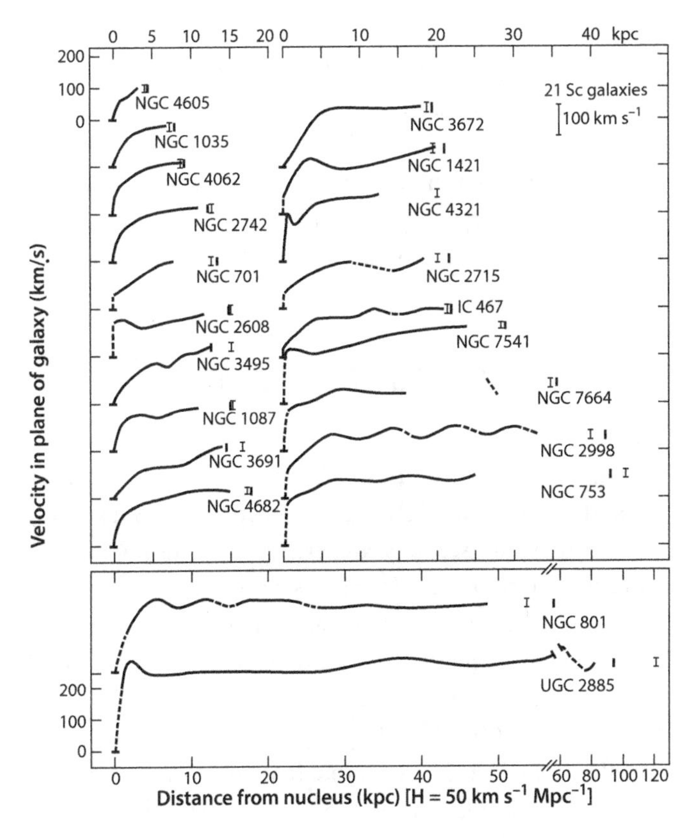

Figure 2.4. The universality of rotation curves. A 1980 compilation of 21 rotation curves from spiral galaxies—all New General Catalog (NGC) or Uppsala General Catalogue of Galaxies (UGC) numbered. All show increasing or constant velocities at large radii, with typical rotation velocities of 100 to 250 km/s.
Source: Rubin, Ford, and Thonnard, *Astrophysical Journal*, 238:471–487, 1980.

quickly collect in a bar-shaped structure in the center of the galaxy, without the extended disk of stars observed in many galaxies (Fig. 2.5). Some observed galaxies do indeed appear not to have a disk and instead display a "bar" of stars at their center, but there are not many of them. Dark matter in a 10-to-1 ratio with normal matter was essential for these simulations to produce a result that looked like most galaxies.

2.4 GRAVITATIONAL LENSING

Matter and dark matter make gravitational fields that change the way objects move. The rotation curves of galaxies and the motion of galaxies in galactic clusters discussed in the previous sections are examples of using massive objects that emit light (such as stars or galaxies) to trace their motion in the gravitational fields and thus determine how much total matter is creating the gravitational field.

A gravitational field created by matter and dark matter can change the motion of light propagating in the field—an effect known as **gravitational lensing**. Astronomers can measure the change in the motion of light to learn about the matter and dark matter that created those gravitational fields. We are used to thinking of gravity working on massive objects like stars and planets, so how can this work if a particle of light, the photon, is massless?

In a gravitational field—for example, near the surface of the Sun—space and time are changed in such a way that clocks tick slower, meaning there is a longer time interval between wave crests of the propagating wave, and objects are compressed along the radial axis to the center of the Sun (Fig. 2.6A). If one thinks about a light wave with crests

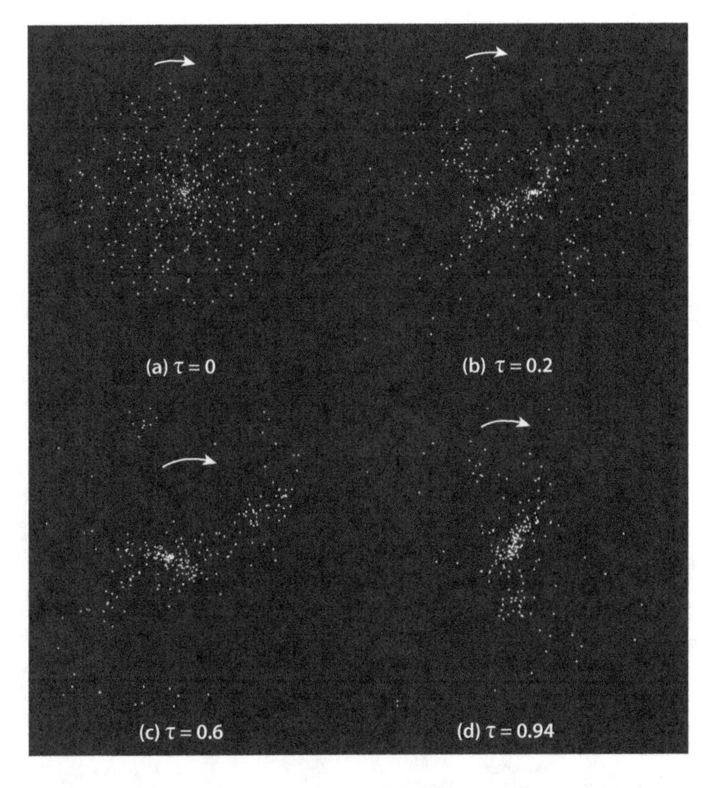

Figure 2.5. Numerical simulation of the motion of stars in a galaxy without dark matter. τ is the number of orbits around the center that a typical particle has completed. By the end of less than one orbit, the stars have collapsed to a bar shape in the middle. Such "barred" galaxies are commonly observed in the universe, but they also have the more familiar spiral arms of stars.
Source: Ostriker, J. and J. Peebles, *Astrophysical Journal* 186:467–480, 1973.

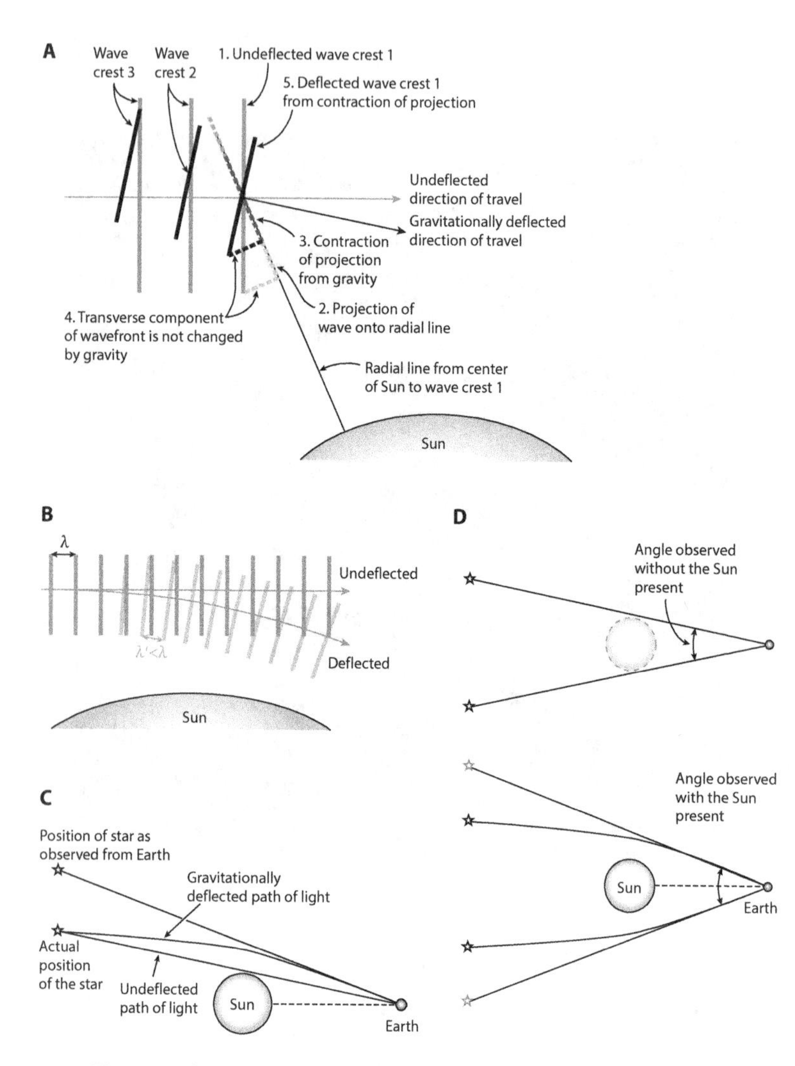

Figure 2.6. How gravity deflects light rays. A) The part of the light wave along the line connecting the wave and the Sun is contracted by gravity, while the perpendicular part is unchanged, tilting the direction that the wave travels toward the Sun. Gravity also stretches out time, so the crests of the wave are farther apart. B) Bending the direction of travel of the light wave toward the Sun. C) The net effect of gravity on the path of the light near the Sun is

and troughs, as shown in Figure 2.6 the contraction in the radial direction tips the light wave toward the Sun, and the slower passage of time shortens the wavelength, resulting in the deflection of the light's direction of travel. For our sun, the deflection angle is about 1/100,000 of a degree for light paths that pass close to the Sun's surface.

Einstein predicted the lensing effect in 1911, and in 1919, Arthur Eddington and others observed it during a total solar eclipse. Eddington measured the angle between two stars, one on each side of the Sun, and compared the angle with the angle between the stars measured during a time of the year when the Sun was not present (Fig. 2.6D). Measuring the angle to 1/100,000 of a degree meant that the measurement had to be made during a total solar eclipse—when the Moon completely covers the Sun and the Sun's glare does not obscure the stars.

Eddington carried out his observations in Sobral, Brazil, during the total solar eclipse, with another expedition on the island of Principe, West Africa. Both teams observed stellar deflection at the same time, and their results agreed, but with large uncertainties. Measurement of this deflection was one of the first indications that Einstein's theory of gravity was correct. This effect has been used many times

Figure 2.6 (continued). to curve the path of the light wave. An astronomer on Earth measures the direction of the light at the Earth, making the star appear to be in a different place in the sky. D) In 1917, Arthur Eddington and others measured the angle between two stars during a total solar eclipse, when the stars appeared on either side of the Sun. The angle they measured was larger than the angle that was measured at other times of the year when the Sun did not appear near the two stars.

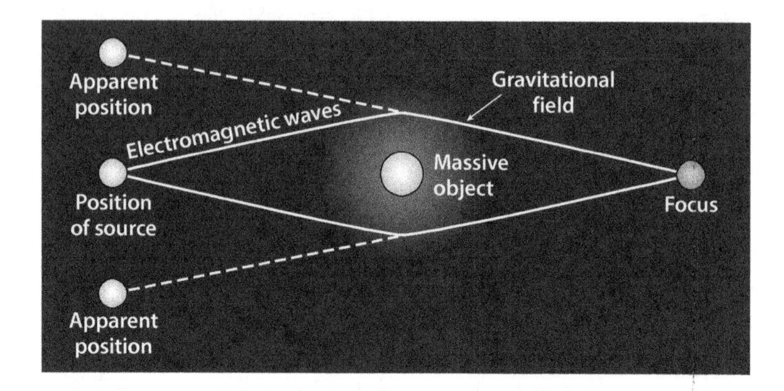

Figure 2.7. How gravitational lensing can result in multiple images of the same object. A massive object (such as a galaxy or cluster of galaxies) in the center of the image bends light from a light source at the left to an observer at the focus, producing multiple distorted images of the source.
Source: Geophysical Fluid Dynamics Laboratory.

since to learn about the way light propagates in gravitational fields as a test of Einstein's theory. It has also been used to determine the mass of galaxies or clusters of galaxies by using the lens effect to measure the total mass of a galaxy or cluster and comparing the results to the light output of the galaxy or cluster, which is related to the mass of the normal matter. The difference between the two measurements is the mass of the dark matter.

If a bright light source, like a **quasar**, lies behind a galaxy, the light from the quasar must pass through the gravitational field of the galaxy to reach an observer on Earth (Fig. 2.7). During the 1990s, an automated survey telescope, the Sloan Digital Sky Survey, found hundreds of thousands of quasars across the sky, and astronomers have

Figure 2.8. Five images of one quasar. The center image is a lensing galaxy between the Earth and a quasar. Acting as a lens of the quasar's image, the lensing galaxy has split the quasar into the four additional images shown.
Source: National Aeronautics and Space Administration (NASA), European Space Agency, and Space Telescope Science Institute.

found all kinds of ways that the gravitational field of a galaxy can distort the image of a bright quasar.

One example of a distorted image that occurs if the quasar lies just off axis between the Earth and the galaxy is known as an Einstein cross, shown in Figure 2.8, where the image of the quasar is split by the light propagating through the gravitational fields of the galaxy into five images of the quasar, each with different brightness.

Measuring the position and brightness of the five images of the quasar gives a measurement of the total mass of the lensing galaxy. Comparing this mass with the luminous mass of the lensing galaxy has shown that there is about 10 times more dark matter than normal matter in the galaxy, roughly consistent with the measurements of the motion of galaxies in clusters and galactic rotation curves. The lensing measurements give a third measurement of the dark matter mass fraction in galaxies, since lensing relies on the effect that gravity has on light.

A second result comes out of the lensing measurements. If dark matter were concentrated in massive objects, such as very large planets, **white dwarfs**, or black holes, the small lensing of these objects would collectively change the amount of light in each image, a phenomenon known as **microlensing**. Measurements of the brightness in each quasar image do not show microlensing, implying that any dark matter clumps in these objects must have a mass less than that of our sun. The microlensing result excludes the idea that most dark matter in a galaxy could be concentrated in supermassive black holes weighing millions of solar masses.

2.5 1E 0657-56 AND THE BULLET CLUSTER

In the 1980s, modifications to the equations describing gravity, collectively referred to as **modified Newtonian dynamics (MOND)**, were developed to explain the measured galactic rotation curves without requiring the presence of dark matter. MOND changes the strength of gravity for galactic distances to accommodate the rotation measurements while keeping gravity's strength at shorter distance unchanged. **Weak lensing** measurements, described below, have used the effect of gravity on the trajectory of light to distinguish between MOND and theories with dark matter. Measurements of two colliding galaxies, collectively called 1E 0657-56,[10] provide a striking example

10. The catalogue number 1E 0657-56 refers to two clusters of galaxies that passed through each other 150 My ago. Two clusters of galaxies nearby also form a cluster of galaxies; and the smaller of the two is known as the **Bullet Cluster**, because it resembles a bullet from a gun. In this section, "cluster" refers to one or the other of the two clusters of galaxies making up 1E 0657-56.

of how weak lensing may be used to distinguish between the influence of dark matter and MOND.

The system 1E 0657-56 lies 3.7 Glt-yr from Earth and consists of two clusters of galaxies—totaling about 40 galaxies between the two clusters—that passed through each other 150 My ago (Fig. 2.9). The two clusters span about 5 Mlt-yr, are now about 500 klt-yr apart, and are moving away from each other at 4,000 km/s. The 1E 0657-56 system has a mass of 2.5×10^{14} solar masses consisting of 1% stars, 11% intracluster gas (mostly hydrogen), and 88% dark matter. The hot intracluster gas emits x-rays that can be detected by x-ray telescopes orbiting the Earth and used to map the location of the intracluster gas. The three components—stars, dark matter (or MOND), and intracluster gas—were subject to different forces during the collision. The reactions of the stars, dark matter (or MOND), and intracluster gas can be used to tell whether there is in fact dark matter in the clusters or the effects are from MOND.

The 1E 0657-56 system lies in front of hundreds of galaxies, and the light from these background galaxies must pass through or near the system to reach Earth. Like starlight passing near the Sun, the light from the galaxies behind the system deflects as it passes through and near the cluster; and the gravity distorts the shape of the galaxies as viewed from Earth, stretching the image horizontally and compressing the image radially (Fig. 2.10). Thus, a circular light source appears as an ellipse, with the smaller axis pointing toward the center of the Bullet Cluster (Fig. 2.10B), an effect called weak lensing. Measuring the weak lensing of hundreds of background galaxies provides the total mass in the system, as well as the position

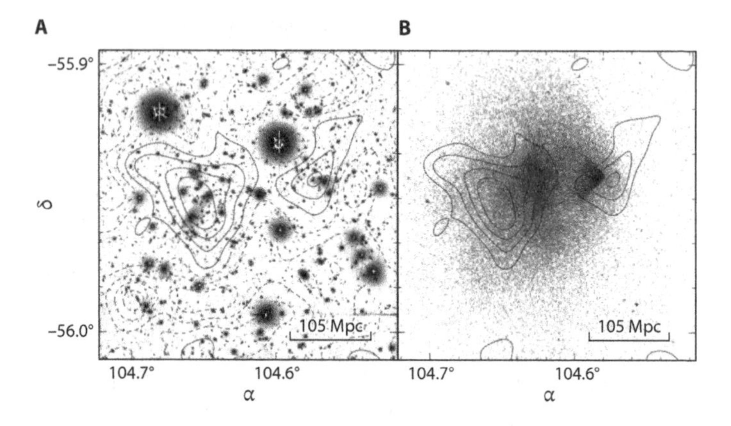

C

Figure 2.9. 1E 0657-56: two colliding clusters of galaxies. A) Mass contours of 1E 0657-56 superimposed on optical image of the two colliding galaxies. The mass coontours, determined from weak lensing measurements, show the highest mass concentrations in the smallest closed curves (left and right of the center of the image). B) The x-ray image of 1E 0657-56 as gray pixels with the mass

of the center of mass in each of the two colliding clusters. The stars, dark matter, and intracluster gas contribute to the total mass in the two colliding[11] clusters that make up the Bullet Cluster.

During the collision, gravity acted in the same way on the stars, putative dark matter, and intracluster gas in each cluster as the clusters passed through each other. The intracluster gas experienced an additional force: The gas in each cluster is hot, allowing the gas in one cluster to interact with the gas in the other cluster via the electromagnetic force through the emission and absorption of light. The exchange of light between the gas in each cluster caused a drag force during the collision, slowing the intracluster gas relative to the stars and dark matter in the same cluster. As a result, the dark matter and stars ran ahead of the colliding intracluster gas, resulting in a separation between the stars and dark matter and gas in each cluster.

The separation was observed by comparing measurements of the position of the intracluster gas and the total

Figure 2.9 (continued). contours from panel A. The colliding normal matter gas in the two galaxies creates the x-rays, and the darkest pixels show the high normal mass concentrations. The highest total mass concentrations have separated from the highest normal mass concentrations, showing that the dark matter, most of the mass in the system, has run ahead of the mormal matter. C) Optical image of 1E 0657-56.

---

11. It is customary to speak of the two clusters in 1E 0657-56 as "colliding." The galaxies in each cluster did not come into physical contact during the collision. The gravity of one cluster caused the galaxies in the other cluster to deflect. The two clusters passed through each other and were disrupted by these deflections, but the clusters were not destroyed by the passage or collision.

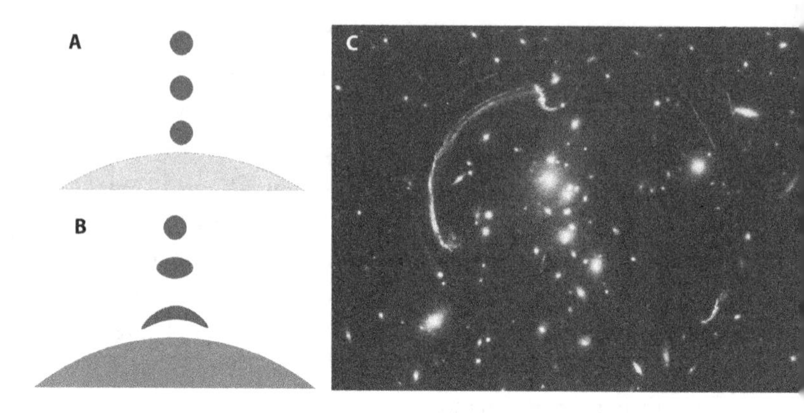

Figure 2.10. Strong gravitational lensing. A) Undistorted images of three distant circular objects. B) When the light rays from the distant objects in A pass close to a massive body, shown as a crescent, the gravitational field stretches the image along the azimuthal direction and compresses the light rays along the radial direction. The distortion is approximately equal to the distance from the lensing object. C) An extreme example of gravitational distortion. This Hubble Space Telescope photo from 2012 shows a cluster of galaxies, called RCS2 032727-132623, about 5 Glt-yr away, and several distorted images (in terms of a clock, the images lie at 12:00, 3:30, 5:00, and between 8:30 and 11:00) that result from strong lensing of a galaxy behind RCS2 032727-132623.
Sources: NASA, ESA, J. Rigby (NASA GSFC), K. Sharon (KICP, U Chicago), and M. Gladders and E. Wuyts (U Chicago).

mass of each cluster. The location of the total mass was found using weak lensing of the hundreds of background galaxies behind and around the clusters to triangulate the position of the total mass in each cluster. The x-ray image of each cluster gave the position of the intracluster gas in each.

The measurement of the x-ray emission was consistent with the expectations for 11% mass in intracluster gas for a cluster of the size of the system; and the lensing measurements showed the stars and dark matter, 89% of each cluster's mass, had separated from the intracluster gas by about 700 klt-yr as a result of the electromagnetic interaction between the gas in the two clusters. Together, these results show that the two galaxies have about six times as much dark matter as normal matter that became separated during the collision.

Astronomers have observed other colliding clusters of galaxies (for example, MACS J0025.4.1222) that show a separation between the dark matter and stars and the intercluster plasma. The results from these observations support the conclusion from the measurements of dark matter in the Bullet Cluster.

There is no evidence of a MOND-type effect that changed the strength of gravity. Observations of other colliding clusters give similar results, but the Bullet Cluster is the most spectacular confirmation of dark matter. By chance, we see 1E 0657-56 from the side where the separation between the gas and stars and dark matter is most apparent.

2.6 LIGHT FROM THE BIG BANG

Hydrogen atoms formed from electrons and protons about 200,000 to 400,000 years after the Big Bang, rendering the universe electrically neutral and allowing light to stream across the universe. The light carries the imprint of quantum mechanical variations in the density of dark matter at the time of hydrogen formation. The light from the Big Bang was first detected in 1963, and an instrument on the COBE satellite produced the first image of the light from the Big Bang in 1991. Since then, a series of ever more precise experiments have sharpened our picture of the early universe and measured the amount of dark matter. This section sketches what happened from the earliest part of the Big Bang until the formation of atomic hydrogen early in the universe.

During its first 10^{-36} seconds, the universe was small enough that light could travel across it in the time the universe had existed. All matter present at that time had a constant average density across the universe; but at these small distances and short times, quantum mechanics played a major role, and there were large spatial variations in the dark matter density called **density fluctuations**. Because the universe was so small, the dark matter could slosh around and smooth out the density differences, erasing them as quickly as they appeared.

Between 10^{-43} and 10^{-32} seconds, the universe went through a **phase transition**, causing a hypothesized expansion of the universe called **inflation**.[12] A phase transition is the rapid conversion of matter from one state to another and, in the case of the very early universe, the phase transition was the release of energy by an early field of particles as they transitioned from a higher to a lower energy state. We do not know much about the particles that underwent this phase transition. They are called **inflatons** and only existed for a tiny fraction of a second in the very beginning of the

12. "Inflation" properly refers to a class of theories that incorporate rapid expansion into the early universe. Theorists believe something like inflation occurred, and observations support this notion, but proving that inflation happened will require very precise measurements of the **cosmic microwave background (CMB)**.

universe. The inflationary phase transition created the precursors to all the particles we know today. The simultaneous creation of all matter drove the expansion of the universe, which doubled in size 86 times before stopping. Having reached a lower energy state, the universe, with all matter and radiation in it, continued to slowly expand and cool.

As the universe cooled, the hot, dense mass of fundamental particles coalesced into protons and neutrons, along with photons and electrons, all at the same temperature. Protons and neutrons formed 1 microsecond after the Big Bang, and electrons and positrons, also created during the phase transition, annihilated with each other, leaving only electrons.[13] At around 1 second, heavier elements—deuterium (the bound state of a proton and a neutron) and helium (the bound state of two protons and one or two neutrons)—began to form, which continued over the next 200 seconds. During inflation, the dark matter density fluctuations were pulled away from each other by the expansion of the universe. The dark matter could not move over to smooth out the fluctuations as it could before inflation, and so the fluctuations became a permanent feature of the universe. The density fluctuations became isolated from each other as the expansion continued. The gravity created by the dark matter perturbations pulled the particles filling the universe into the regions where there was more dark matter and away from the regions where there was less dark matter. The normal matter of the universe now carried the imprint of the original dark matter density fluctuations.

---

13. Only electrons were left because there were more electrons than positrons created after the phase transition. The reason for this imbalance between electrons and positrons remains unexplained.

Each gravitational field created by the density fluctuations with excess dark matter is called a **gravitational well** or potential well. Like a water well, a potential well refers to a place where things fall in and cannot get out. "Potential" is short for "gravitational potential," meaning the wells result from the gravity created by fluctuations with more dark matter than average.

At around 200 seconds, the temperature of the universe was about a billion degrees. The protons, electrons, and photons all interacted via the electromagnetic interaction: the electrons and protons attracting each other through the Coulomb interaction; and the photons interacting by bouncing off the charged electrons and protons, a process called **scattering**. At this time, the rapid motion caused by the thermal energy from the billion-degree temperature of the universe prevented the electrons and protons from binding to form hydrogen.

Once they reached the same temperature, the protons and electrons fell into the dark matter potential wells under the force of gravity (Fig. 2.11). As the electrons and protons converted gravitational energy into light by emitting photons, the photons scattered off the charged electrons and protons and were dragged along with them. This gas of electrons, photons, and protons fell toward the center of the well, with more and more photons being created from the energy loss of the charged particles, raising the temperature and pressure of the gas in the well. The infall continued until the pressure of the photons scattering off the electrons and protons, which pushed the gas out of the well, overcame the gravitational pressure pushing the gas in.

At that point, the gas stopped falling into the potential wells and began expanding under the pressure of the photon scattering. The expansion continued until the gas returned to about the size it had when it began to fall into the potential well. Then the expansion stopped, and a new contraction began. The gas in each potential well followed this contraction-expansion-contraction cycle, with the time for each cycle proportional to the size of the potential well (smaller potential wells had shorter contraction and expansion times).

Inside a potential well, while the gas was contracting and expanding, the dark matter whose gravity created the potential well was also falling in and out of the well. However, the dark matter did not interact with any other matter in the potential well, except by gravity, so it could not lose energy and stay in the middle of the well. All the dark matter could do was fall through the center of the well and out the other side, sort of like a skateboarder skating back and forth, up and down the sides of an empty swimming pool. While the dark matter may have sloshed back and forth through the potential well, the overall distribution of dark matter remained the same; and since there was so much more dark matter than normal matter in the potential well, the gravity making up the well stayed about the same during the expansion and contraction cycles of the normal matter.

The universe continued to expand, causing the dark matter potential wells to move farther apart. The potential wells lost the ability to share matter at the end of inflation and became isolated structures, with the electron-proton-photon gas contracting and expanding over periods proportional to the size of the well.

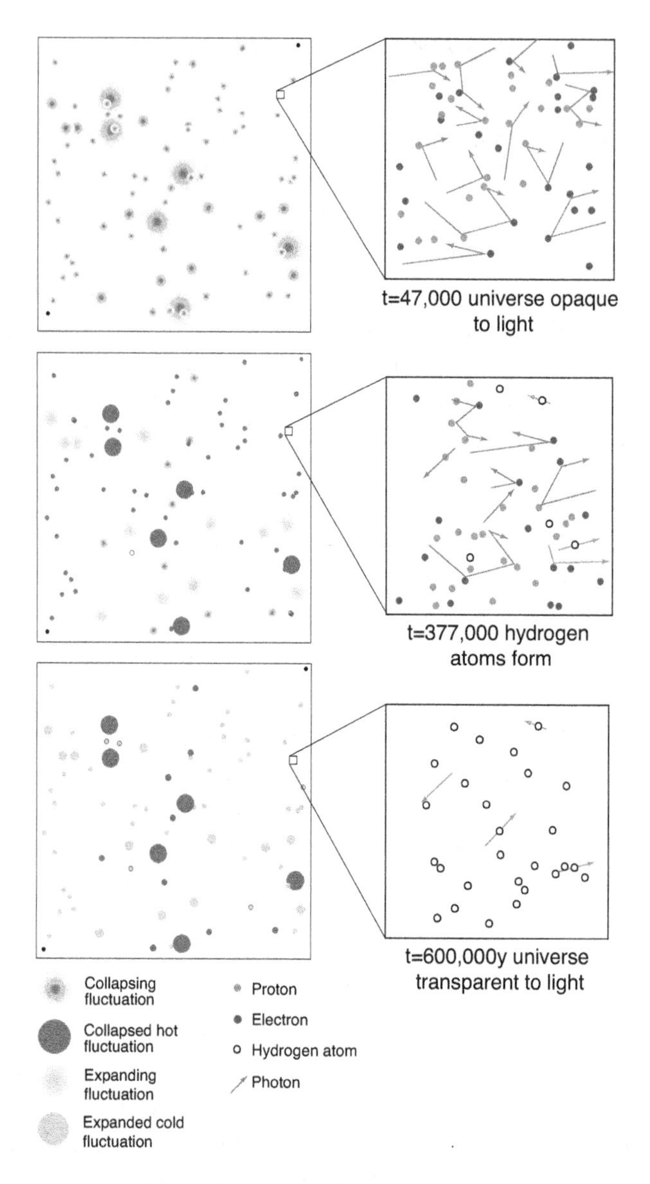

Figure 2.11. Time sequence of the release of light from the Big Bang. The three panels on the left show a patch of the early universe at $t = 47{,}000$, 377,000, and 600,000 yr after the Big Bang. The gray

The expansion-contraction cycles went on for about 200,000 years, by which time the universe had expanded to 1/1,100 the size of the universe today and had cooled to a temperature of 3,000 K—cool enough that the electrons could bind with the protons, forming hydrogen atoms in ever-increasing numbers. By about 600,000 years after the start of the Big Bang, the universe had cooled enough so that all the electrons and protons were bound to each other in hydrogen atoms.

The number of unbound charges in the universe went to zero. With fewer charged particles in the universe, the photons traveled farther before being scattered by a charged particle. When the electrons and protons were all bound into hydrogen atoms, the universe became transparent to

Figure 2.11 (continued). circles show collapsing and expanding regimes of the early universe caused by dark matter density fluctuations. Each panel to the right shows a microscopic view of the corresponding patch of the early universe, with the arrows showing photons and the circles showing electrons, protons, and hydrogen atoms. The largest fluctuations measure about 1° on the sky today. At 47,000 yr after the Big Bang, the photons, electrons, and protons are in equilibrium, and the fluctuations begin to collapse. The mean free path of the photons at this time is much shorter than the size of the fluctuations, and the photons are "dragged" into the fluctuation with the matter. The fluctuations collapse until the pressure is high enough to counteract gravity, at which time the fluctuations begin to expand. The time for a fluctuation caused by the dark matter to reach maximum compression is proportional to the size of the fluctuation. At 377,000 yr, the universe has cooled to 3,000 K, hydrogen begins to form, and the universe begins to become transparent to light. At this point, the photons stream away from the fluctuations, and they have been propagating ever since.

Figure 2.12. Structure in the CMB. A) The full-sky map of the CMB from COBE in 1992. The COBE differential microwave radiometers (DMRs) could measure the temperature of sky patches as small as

photons, which were free to travel across the universe, as they have done for 13.7 billion years.

Upon the formation of hydrogen, the photons streamed out of the dark matter potential wells. There were no more charged particles in the potential wells to keep them there by scattering; and, without the photons to generate pressure against the gravity pushing the normal matter into the wells, the normal matter continued to fall in the wells, forming much denser regions and triggering star formation. These regions would eventually become galaxy clusters.

The light that streams away 600,000 years after the Big Bang carries the imprint of the dark matter fluctuations (Fig. 2.12). The normal matter that collapsed into a potential well when the hydrogen formed would have released more photons than a region outside a potential well. When those photons reach Earth, they carry with them a map of the surface of the sky 13.7 billion years away, created on average 377,000 years after the Big Bang. At the time of their release, the photons were in the infrared part of the electromagnetic spectrum; but during their long journey to Earth, the universe expanded to 1,100 times its previous size, stretching the photons from infrared to microwaves,

Figure 2.12 (continued). 7°. Structures in the variation of temperature are visible. (Source: https://lambda.gsfc.nasa.gov/product/cobe/). B) The full-sky map from the Wilkinson Microwave Anisotropy Probe (WMAP) that was able to discern structures as small as 0.2°. C) An all-sky map from Planck, where structures in the temperature map on many different scales as small as 0.08° are visible.
Source: ESA and the Planck Collaboration.

with an average frequency of 160 GHz. This radiation is called the Cosmic Microwave Background (CMB).

When viewed from a distance, the CMB photons from a collapsed dark matter potential well appear hotter, and the light from an empty region between collapsed regions appears cooler. The largest hot regions had just collapsed for the first time when the photons were released, and they subtend about 1° on the sky today. The next-largest regions collapsed and expanded once. They are half the size of the largest, subtending about half a degree. The third-largest potential wells collapsed, expanded, and collapsed again, and so are about a third of the size of the largest regions, and so on. By measuring the size and temperature of the different regions, the amount of dark matter in the earlier universe can be inferred from the small temperature variations.

Temperature differences of 300 parts per million make the CMB temperature measurement very difficult. Temperature measurements of such precision require a radiometer—a very sensitive detector of microwaves. Steady improvements in radiometers led to a succession of experiments that have measured the light from the Big Bang over the past 50 years.

A series of spacecraft—the Cosmic Background Explorer (COBE) (1989 to 1993), Wilkinson Microwave Anisotropy Probe (WMAP) (2001 to 2010), and Planck (2009 to 2013)—made all-sky measurements with increasing precision (see Fig. 2.12). During this time, ground and balloon-borne radiometers measured patches of the sky with high spatial precision. The BOOMERanG experiment measured the first evidence for the hot spots from the largest potential wells in 1999. A year later, the MAXIMA balloon experiment measured the first-, second-, and third-largest potential wells. Since then, combined data from many experiments have allowed observation of the ten largest patches. Analysis of the size and temperatures of the patches tells us that the universe today contains 4.9% normal matter, 26.8% dark matter, and 68.3% dark energy; and the age of the universe is 13.8 billion years—all consistent with other measurements.

In the late 1990s and early 2000s, the Lambda-CDM model emerged that combined dark energy, dark matter, and the Big Bang theory to explain most observations of the early universe. "Lambda" refers to the parameter in general relativity that captures the observed effects for dark energy, and "CDM" refers to cold dark matter, described in Chapter 4. At least three Nobel Prizes have recognized the development of the Lambda-CDM model. The Lambda-CDM model is not the final word on either dark matter or dark energy, but it provides a description of what we currently know.

Astronomers have measured the dark matter density and distribution in many other ways. For example, supernova explosions and the distribution of galaxies in galaxy clusters also provide very precise measures of how much dark matter there is. The measurements of dark matter described in this chapter result from measurements of galaxies with sizes of 300 klt-yr, clusters of galaxies with a size of 30 Mlt-yr, to the early universe with a size of 3 Glt-yr. Measurements of objects differing in size by a factor of 10,000 give a consistent result for the density of dark matter in the universe, a remarkable achievement.

What about making measurements on smaller distances? Inside a galaxy, for example, about halfway from the galactic core, where the Earth orbits the Sun, normal matter is 10,000 times denser than the dark matter. For this reason, we can forgive humankind for not being aware of dark matter sooner: After all, astronomers did not understand there were galaxies until 1925, 10 years before Zwicky's first measurements; and the great preponderance of normal matter in a galaxy obscures the subtle gravitational influence of dark matter.

Finding dark matter here on Earth, with so much normal matter around, will require other means (discussed in Chapters 5, 6, and 7). Before we consider that, however, we must first explore what we know about normal matter and why particle physicists believe that dark matter may be particles.

3
NORMAL MATTER: THE STANDARD MODEL

The picture of all material objects being composed of atoms emerged about 120 years ago. Since then, a great deal of work has gone into finding and cataloging the properties of all the particles that make up atoms. Since the 1960s, astronomers and physicists have developed the Big Bang theory and used it to calculate how much of each kind of particle inhabits the universe. In this chapter, we start with a description of all the particles we know about. In Chapter 4, we will consider which known particles could be dark matter—and if there are enough of them to account for the amount of dark matter that astronomers observe.

In the mid-1970s, a series of particle physics experiments worldwide, supported by a great deal of theoretical work, resulted in the Standard Model—a theory of how all known particles interact with one another. We need a basic understanding of the Standard Model and its components to give context to our notion of what dark matter is—and isn't.

3.1 PARTICLES AND INTERACTIONS

We begin with a discussion of the key players—the particles and interactions (forces) that help to define the model.

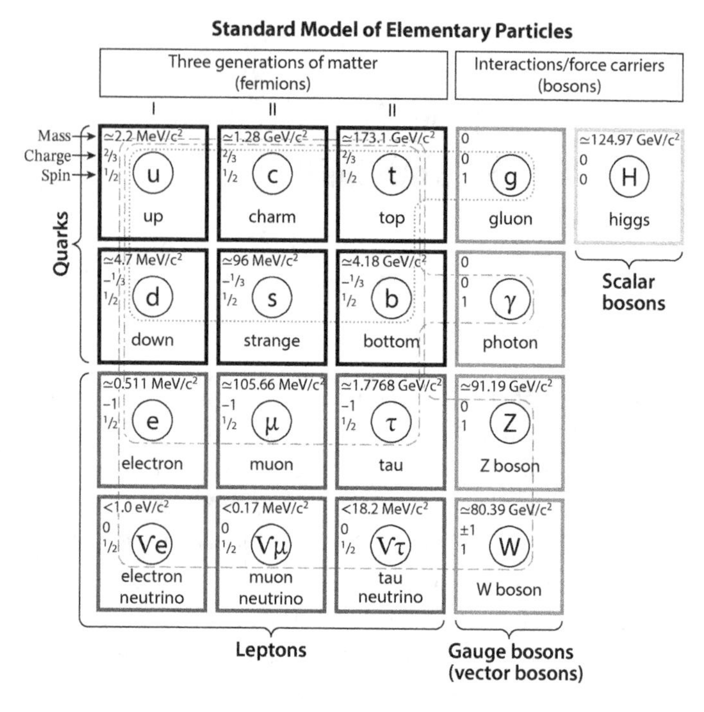

Figure 3.1. The particles of the Standard Model. The dotted line encloses the strongly interacting particles, the dash-dot line encloses the electromagnetically interacting particles, and the dashed line encloses the weakly interacting particles. In the model, the higgs boson is designated with a lowercase *h*; the text of the book uses the more common convention of uppercase *H* for the Higgs boson.

Figure 3.1 shows how the particles and forces are arranged in the model. The first three columns are **fermions** that may bind together to form composite particles, such as **hadrons** or atoms. Each column denotes a **generation** of particles, with column I being the lightest and column III being the heaviest generation. Only the particles in column I are stable. Those in columns II and III decay in a fraction of a second to the particles in column I.

The top two particles in each column are the quarks that bind together to form nuclear matter—hadrons. The quarks on the top row share properties across the three columns, as do the quarks in the second row. As an example, two *u* quarks and a *d* quark—*uud*—form a proton; *udd* is a neutron. A particle formed by three quarks is called a **baryon**.

The electron occupies the third row of column I. The muon and **tau** are also in row 3 in columns II and III, respectively. Both are unstable and quickly decay to electrons, emitting **neutrinos** and maybe some nuclear particles. An electron orbiting a proton makes a hydrogen atom; protons and neutrons may bind together to make a nucleus, and if you add one orbiting electron for each proton, chemical elements result. Each particle in row 3 has an uncharged partner in row 4 called a neutrino, and we shall see that neutrinos do not orbit other particles to form more complex particles. The particles in rows 3 and 4 are collectively called leptons.

The **force carriers** lie in the fourth column of Figure 3.1. Three forces appear in the Standard Model: the strong, weak, and electromagnetic. The force carriers fly between the quarks and leptons to make them "feel" forces from each other. For example, a photon may travel from a quark in a proton to an electron, pulling the electron toward the proton, making the electron orbit the proton. From top to bottom, the forces carriers are the gluon, for the strong force; the photon, for the electromagnetic interaction; and the *Z* and *W* for the weak interaction. Another term for force carriers is **gauge bosons**.

Finally, in the fifth column is the **Higgs boson**, called a **scalar boson**. The Higgs particle, the last particle to be observed (in 2012), has unusual properties that we consider later in the chapter.

When first proposed in the early 1970s, the Standard Model contained four kinds of quarks, two kinds of electrons (the electron and the muon), two kinds of neutrinos, five force carriers, and predicted the Higgs boson. Over time, as a result of experimental observation of new quarks and leptons, the Standard Model has been extended to include two more kinds of quarks, one more electron, one more kind of neutrino, and massive neutrinos, but it remains structurally unchanged. We can now say that the Standard Model is complete, in that there do not seem to be any missing particles or interactions. The Standard Model has made sensible predictions of physical phenomena that have been experimentally confirmed. A product of over 40 years of work by several thousand scientists, the Standard Model represents a remarkable achievement.

No particle or force in the Standard Model explains the observations described in Chapter 2 that are attributed to what we call dark matter. As we shall see, none of the particles in Figure 3.1 has the properties needed for dark matter, and there is no "slot" in the model for a new dark matter particle. Suppose you were a coin collector, with all the slots in your coin books filled. One day, you hear of a coin of a date and denomination combination that is not in your book. What do you do? Our questions do not stop there, though.

The Standard Model does not include gravity, which stands as a completely separate, well-tested theory, and the Standard Model has some odd features that seem unlikely. The first such feature relates to the Higgs boson (and we will discuss it separately in Section 3.2). The second is that the Standard Model predicts that the strong interaction that binds the nucleus should have certain properties that have not yet been observed (we will discuss this issue in Chapter 7). Theorists have worked hard at developing extensions to the Standard Model that explain these two oddities. This section starts with a description of the Standard Model and then turns to the remedies for these deficiencies, which touch upon dark matter. Each remedy has implications for dark matter that will be discussed later in the book.

There are two types of particles in the Standard Model—fermions and bosons—that have different quantum mechanical properties. The particles that make up the nucleus (the quarks) and the particles like the electron (the leptons) are all fermions. Fermions have nonzero masses, and the Standard Model predicts that the total number of fermions in the universe remains constant (i.e., if a process absorbs a fermion, then the process must emit a fermion, or two fermions and an anti-fermion, and so on). The Standard Model also predicts that quarks cannot exist in isolation. They must always be bound with other quarks or anti-quarks into particles called hadrons. The particles that we are most familiar with, the proton and neutron, are bound states of three quarks. Isotopes are bound states of protons and neutrons, and the elements are isotopes with electrons orbiting them.

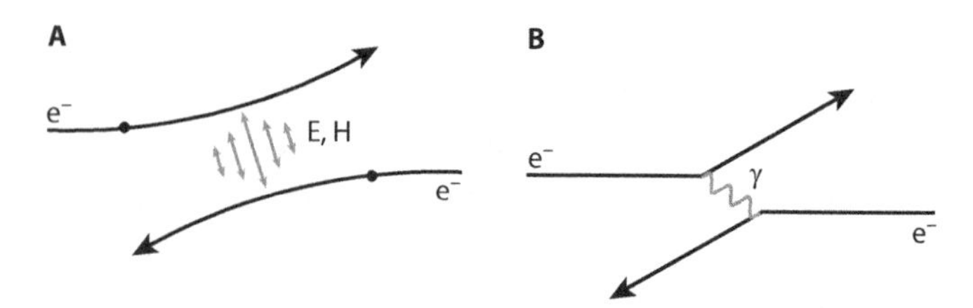

Figure 3.2. Two electrons, shown as dots, interacting in the classical and quantum pictures. The two electrons collide nearly head-on. A) In the classical view, the electric (E) and magnetic (H) fields push the electrons apart, altering their trajectories. B) In the quantum view, the electrons exchange a packet of electromagnetic energy called a photon (γ), deflecting the trajectories in much the same way as in the classical view.

The bosons carry forces between the fermions. For example, an electron flies toward another electron and is deflected or scattered by the electric forces acting between them (Fig. 3.2). The quantum mechanical picture of the same process is that one electron emits a photon, deflecting that electron's direction of travel. The photon travels to the other electron and is absorbed, deflecting the second electron's trajectory. We say that the electrons have exchanged a photon. All of the bosons, except for the Higgs boson, are called vector bosons and serve to transmit forces as described above (Fig. 3.3). Unlike the fermions, the number of bosons in the universe is not constant.

3.2 THE HIGGS BOSON

The Higgs boson has a special role in the Standard Model. The leptons, quarks, and vector bosons have the property that if no particle of their kind is present in a region of

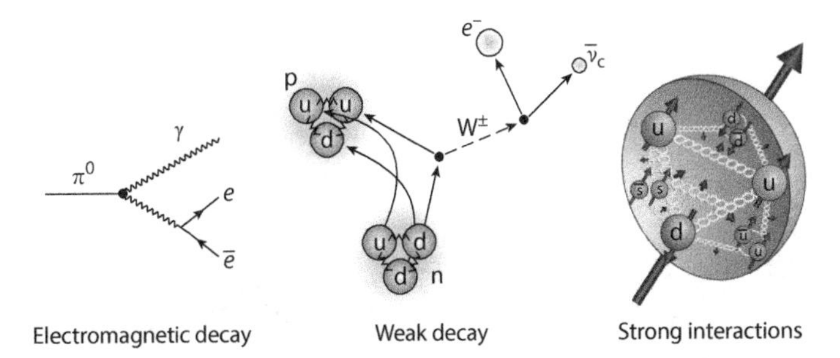

Figure 3.3. Diagrams illustrating three of the four forces. Left: The electromagnetic decay of a particle, the neutral pion, to two photons (indicated by the wavy lines). Middle: Beta decay of a neutron to a proton. The W boson, indicated by a dashed line, carries the weak force. Right: The quarks, shown as spheres, that make up the proton are bound together by the exchange of gluons, carriers of the strong force, shown as chains.
Left: From P. Adlarson, W. Augustyniak, et al. "Search for a dark photon in the $\pi 0 \rightarrow e + e - \gamma$ decay" (*Physics Letters* B, 726: Issues 1–3, 2013, Pages 187–193); Middle: From "Neutron Beta Decay Asymmetry Nab" (ORNL Physics Division, UT-Battelle for the Department of Energy); Right: From "Spun out of proportion: The proton Spin Crisis" (Quantum Diaries © 2021 Interactions.org).

space and time, we can use the Standard Model to compute zero probability of that particle being present in that region. If a particle enters or is created in that region, the probability for the presence of that species of particle will be nonzero (Fig. 3.4). For the Higgs boson, in contrast, the Standard Model predicts a constant value for its presence, even if no Higgs particle is present: The probability of the Higgs being present is the same everywhere in space and time. The Higgs is everywhere, all at once, and this presence is called the **vacuum expectation value (VEV)**. The

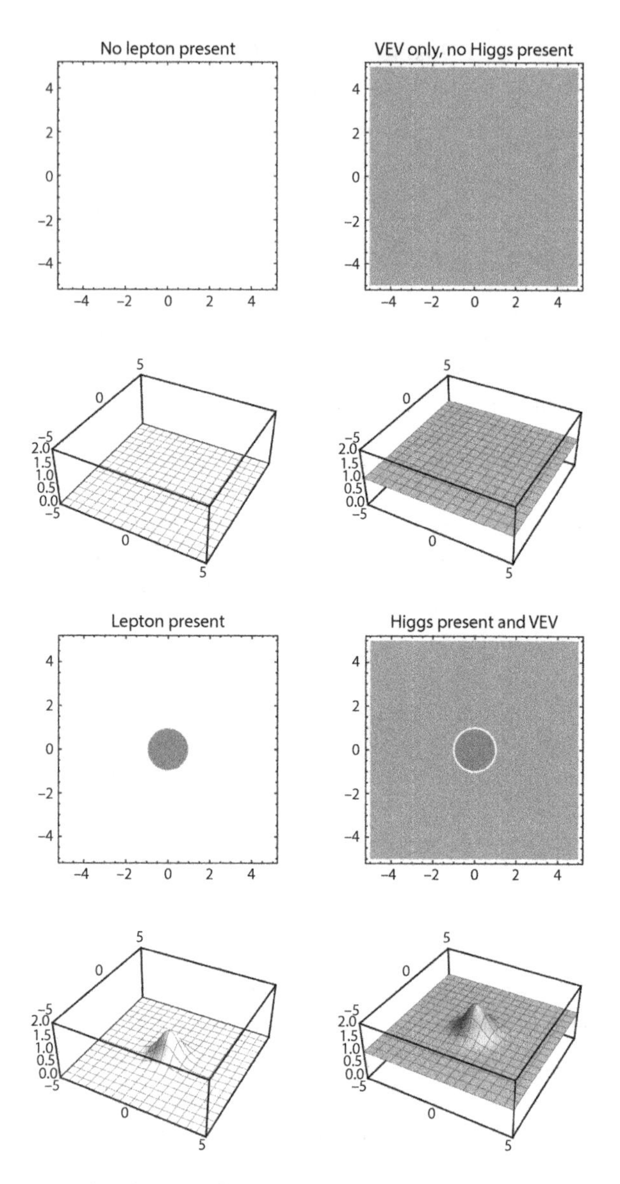

Figure 3.4. Schematic illustration of vacuum expectation value. The top row of figures shows a two-dimensional region of space without and with a particle present, indicated by a disk. The lower row of figures shows the probability distribution at each point in space.

interaction between the Higgs boson's vacuum expectation value and the Higgs boson particle gives all the particles their masses. If Higgs particles are present, the Standard Model predicts their probability of presence in addition to the vacuum expectation value.

The Standard Model specifies all the ways that fermions and bosons interact with each other; and, for over 30 years, almost all of the predictions of the Standard Model have agreed with actual measurements. For example, a series of measurements in the 1990s enabled the Standard Model's prediction of the mass of the Higgs boson at 134 proton masses, which was borne out with the discovery of the Higgs in 2012.

3.3 TESTING THE STANDARD MODEL

From time to time, measurements disagree with Standard Model predictions. However, except for one prediction, subsequent and better measurements have agreed with Standard Model predictions. Right now, the one measurement that does not agree is the Standard Model's prediction that a heavy electron, called the muon, will spin in a magnetic field; and the Standard Model gives the spin rate with a precision of about half a part per million. However, an experiment at Brookhaven National Laboratory in New York measured the muon spin frequency and found a difference by two parts per million from the Standard Model prediction, with four times the accuracy of the calculation. The discrepancy between theory and experiment led to ideas that there may be a deficiency in the theory and that the Standard Model will need to be changed somehow. An experiment at Fermi National Laboratory in Illinois repeated the measurement with improved precision, collecting data from January 2017 to 2020 and achieving better precision than 1 part per million. Both measurements agree with each other and disagree with the Standard Model prediction. The experiments and prediction differ enough to be very interesting but not enough to be decisive. The Fermilab experiment will continue to collect and analyze data, and several theory groups are working on different approaches to their predictions. Stay tuned.

Despite success in all but one of its predictions, the Standard Model has some other problems. The first is that the model shows that the electromagnetic, weak, and strong forces—three of the four known forces—arise from the idea of fermions interacting with each other by exchanging bosons. Einstein's theory of gravity acts on particles by bending space and time, altering their trajectories. Aligning gravity's bending of space and time with the Standard Model picture of interaction via boson exchange has proven difficult. String theories are an attempt to combine the Standard Model with gravity, but so far they have not yielded a model with testable predictions.

Second, the Standard Model does not have a particle that fills the role of dark matter, which we will discuss in Chapter 4.

Third, the Standard Model predicts that the Higgs boson will have a much larger mass than has been observed, unless the parameters describing the Higgs boson have very specific values—a problem referred to as the **fine–tuning problem**, because the parameter's values must be fine-tuned to get the desired result.

A fourth problem is a feature of the strong interaction that holds quarks inside nucleons, and nucleons in the nucleus. Measurements indicate that all strong interactions are time-symmetric, meaning if you film any interaction and then watch the film backward, the same laws govern the backward viewing as the forward viewing. However, the Standard Model allows (but does not require) strong interactions that are not time-symmetric. Many theorists believe that an undiscovered particle, the axion, prevents time-asymmetric interactions in the strong interaction. Axions would weigh less than a trillionth of a proton and would interact with electric and magnetic fields. We will discuss how they might be detected in Chapter 7.

These four problems are not failures of the Standard Model. Instead, theorists view them as fruitful places to think about extending the Standard Model so it explains more of our observations. Over the past 40 years, for example, a major theoretical effort has gone into the theory of **Supersymmetry (SUSY)**. The idea behind SUSY is that every fermion has an as-yet-undiscovered boson partner, and every boson has an undiscovered fermion partner. The math works out that the Higgs boson mass has the observed value for a wide variety of SUSY parameter values, removing the fine-tuning problem. Some SUSY partners could be dark matter particles, typically weighing about a hundred proton masses. The particles of SUSY would need to be adjusted to ensure that the new dark matter particle does not interact with normal matter, which led to the name Weakly Interactive Massive Particle—WIMP—for the new particle. Though predicted by SUSY, "WIMP" has come to be a generic term for any massive particle that fills the role of dark matter. While promising, no experiment has so far supported the SUSY idea, though experiments continue to look for evidence for both WIMPs and other SUSY partners.

Theorists have also attempted classical solutions by modifying Newton's laws in such a way as to reproduce the galactic rotation curves described in Chapter 2 without resorting to dark matter. MOND keeps Newton's laws the same for distances smaller than a galaxy but makes gravity a little stronger for distances larger than a galaxy in an effort to explain why gas and stars observed near the fringes of galaxies appear to move faster than Newtons' law predicts. MOND theories also require changes to Einstein's well-tested gravity theory and do not predict, for example, the separation between normal and dark matter observed in 1E 0657-56, as described in Chapter 2.

There have been many other ideas and modifications to gravity or the Standard Model to explain the observations we ascribe to dark matter. In Chapter 4, we will look at why none of the Standard Model particles can be dark matter. Chapters 5 through 7 then will describe the experiments that seek physics beyond the Standard Model that can explain dark matter.

4
WHAT DARK MATTER IS NOT

A physicist working in the 1950s, about 20 years after Zwicky's paper, would have little difficulty coming up with theories about the missing mass in the Coma Cluster to explain Zwicky's observations. Low-mass stars, planets, or large clouds of gaseous hydrogen could have been the dark matter needed to explain the high speeds of the galaxies in the Coma Cluster. Neutrinos emitted by nuclear beta decay in a fission reactor had just been observed in 1956 by Frederick Reines and Clyde Cowan. Neutrinos of this energy could penetrate a string of 200 Earths, so it would be easy to explain why dark matter did not interact with particle detectors on Earth if dark matter were massive neutrinos. Dark matter could be heavier elements, neutron stars, or black holes. In the late 1950s, nuclear physics was just emerging, and astrophysics did not yet have the technology for making precise measurements in space, so dark matter could have been just about anything. As time passed, astronomy and particle and nuclear physics advanced; and by the early 1980s, it became clear that dark matter could not be explained by any known particle.

This chapter explains why dark matter cannot be any of the particles in the Standard Model. Section 4.1 explains how the Big Bang created all the particles in the Standard Model, showing that not enough protons were created to account for the effects of dark matter. Section 4.2 explains the origins of neutrinos from the Big Bang, which are also too few to explain dark matter. Section 4.3 considers compact objects—neutron stars, black holes produced since the Big Bang, and failed stars—and the observations that showed they cannot be dark matter. Finally, Section 4.4 describes the proposed modifications to the theory of gravity made to explain dark matter and the observations that show that this idea, too, cannot explain dark matter.

Black holes need some mention here (and we will say more about them in Chapter 6). Black holes may form in three different ways. In today's universe, the remnant of an exploding star—a supernova—may collapse into a black hole, a black hole may merge with a neutron star to form a larger black hole, or two neutron stars may merge into a black hole. In all these scenarios, the resulting black hole was created by normal matter produced in the Big Bang. Black holes also may have been produced before the Big Bang, as Section 4.3.2 describes.

4.1 MAKING VISIBLE MATTER: THE BIG BANG

The Big Bang theory was first proposed by Ralph Alpher, Robert Herman, and George Gamow in 1947 and later developed in the 1960s, leading to the prediction that the light released when neutral hydrogen formed would redshift to radio frequencies that would be observable today as the CMB. The observation of the CMB at a temperature of 2.7 K in 1963 established the Big Bang picture, and the early predictions and measurement of the element abundances—starting in the 1970s—began to reveal that there was a "dark matter problem."[14]

The emergence of a more precise Big Bang theory in the late 1970s, supported by advances in particle physics, began a process of elimination of dark matter candidates from known particles. Within a decade, the idea that dark matter must be a new kind of particle began to emerge. Big Bang theory applies Einstein's General Theory of Relativity to the universe as a whole, as was started by Howard Robertson, Arthur Walker, Alexander Friedmann, and Georges LeMaître in the mid-1920s, and the theory drove forward the experimental and theoretical advances in nuclear physics after World War II.

We want to know whether enough protons were created in the Big Bang to account for the amount of matter in the universe measured in the astronomical measurements described in Chapter 3. If enough protons were made in the Big Bang to explain the effects of dark matter, then we have to look for nuclear matter somewhere besides stars and galactic dust. However, if the Big Bang did not make enough protons to account for dark matter, then we will have to find ways that the Big Bang could have made

14. The uniformity of the CMB presented a major problem until the inflationary theory appeared in the early 1980s. Although there is no direct evidence for inflation (the initial rapid expansion of the universe), the idea fits so well that most cosmologists believe inflation (or a variant of it) explains the uniformity of CMB temperatures in the cosmos. Further measurements of the CMB in the next few years could greatly strengthen the evidence for inflation. In spring 2014, the BICEP2 experiment released measurements that strongly supported the inflationary picture of our universe. However, further analysis showed that the results may have been contaminated by spurious signals from dust in the Milky Way.

something else that can account for the effects of dark matter.

According to inflationary theory (as yet unproven), when inflation ended at 10^{-32} seconds after the Big Bang, unknown primordial particles carried much of the energy of the universe. The primordial particles must have been very massive—millions or more times the mass of the proton. As the universe expanded, the primordial particles decayed into the particles of the Standard Model. By the time the universe had expanded and cooled to a temperature of 10^{15} K, 10^{-12} s after the Big Bang started, the primordial particles were all gone, and the universe was a very hot gas of their decay products (i.e., Standard Model particles: quarks, leptons, neutrinos, photons, and gluons).

At this point, we will take a little detour and consider gases of particles that interact with one another. Because the universe spent its first few million years as a gas with interacting particles, it is worth understanding something about such a gas.

A gas is a system of unbound particles, and the average kinetic energy of each particle is proportional to the temperature of the gas, assuming the temperature does not change too rapidly. What is "too rapidly"? Suppose we have a gas with type a particles and type b particles, and a particles can annihilate to make b particles and vice versa:

$$a + a \leftrightarrow b + b. \tag{4.1}$$

Suppose the a particles have half the mass of the b particles; a will need enough kinetic energy to convert to the mass of the b particles. How much energy? From relativity, $E = mc^2$, so if the b particles have mass m, and a particles have mass $m/2$, each a particle will need $mc^2/2$ of kinetic energy to make two b particles. The temperature gives the average kinetic energy of a particle, so if the temperature corresponds to an energy of $mc^2/2$, some a particles will have more than $mc^2/2$, and the reaction $a + a \rightarrow b + b$ will happen, though less than at higher temperatures. As the temperature gets lower and lower, fewer and fewer a particles will have enough energy to make b particles, and eventually the reaction will not take place at all.

The reaction $b + b \rightarrow a + a$ can happen at any temperature, so as the temperature goes down and the reaction $a + a \rightarrow b + b$ stops, the reaction $b + b \rightarrow a + a$ continues. More and more b particles disappear, and the gas converts completely to a particles.

The b particles produced in the reaction $a + a \rightarrow b + b$ have, on average, less kinetic energy than implied by the temperature, since on average, the kinetic energy of the a particles has gone into the mass of the b particles when they were created. The particles in a gas bump into one another and share energy; and, with time, the average kinetic energy of a b particle produced by the **particle annihilation** of a particles increases the thermal energy. The gas being in equilibrium means that the temperature of the gas changes more slowly than the time it takes for the particles in the gas to acquire the average thermal energy of the gas.

At 10^{-12} seconds after the Big Bang started, the temperature was 10^{15} K and the universe was composed of quarks, leptons, neutrinos, photons, and gluons, all well-understood particles of the Standard Model. At this point, the temperature was high enough that the particles' energies were large compared to their masses, so all the different species of particles would be in thermal equilibrium with one another. For example, the rules of the Standard Model say a quark and an anti-quark can annihilate to form a pair of photons, and that two photons can fuse to make a quark-anti-quark pair. These reactions may be written as

$$q + \overline{q} \leftrightarrow \gamma + \gamma, \tag{4.2}$$

where q is a quark, $\overline{q}$ is an anti-quark, and γ is a photon. The double-ended arrow indicates that the rate of photon fusion is the same as the rate of anti-quark annihilation, so there is a constant number of photon pairs and $q\overline{q}$ pairs in the universe at each temperature. The quarks and anti-quarks are massive, so they can annihilate to form the massless photons no matter how low the temperature gets (when Eq. 4.2 goes to the right). For the photons to annihilate to quarks and anti-quarks (Eq. 4.2 going to the left), the photons must have enough energy to make the quarks themselves, and any additional energy goes into the kinetic energy of the quarks. Since the temperature of the universe gives the typical energy of the particles in it, there is some temperature at which photons can no longer fuse to make quarks and anti-quarks, and then Equation 4.2 only goes to the right. Thermal equilibrium in the early universe meant reactions like Equation 4.2 went both ways, fixing the number of particles on each side of the equation to a specific value at a specific temperature.

Leptons, electrons, positrons, neutrinos, and anti-neutrinos also come into thermal equilibrium:

$$e^{+} + e^{-} \leftrightarrow \nu + \overline{\nu}, \tag{4.3}$$

and electrons and positrons come into equilibrium with photons:

$$e^+ + e^- \leftrightarrow \gamma + \gamma. \quad (4.4)$$

Even though they do not interact with each other, the neutrinos and photons are in thermal equilibrium, since they can interact through the electrons. The neutrinos are also in equilibrium with the quarks. In today's much cooler universe, the electron-positron annihilation to neutrinos (Eq. 4.3) would be about 10^{10} slower than the electron-positron annihilation to photons. At 1 μs after the Big Bang, the temperature of the universe was still high enough that electron-positron annihilation to neutrinos (a weak process; Eq. 4.3) and photons (an electromagnetic process; Eq. 4.4) had the same strength, keeping electrons, positrons, photons, neutrinos, and their anti-particles in equilibrium with one another.

The universe continued to expand and cool; and before 1 μs, T = 10 GK, and quarks and anti-quarks were in equilibrium with hadrons and anti-hadrons. Between 1 μs and 30 μs, the universe cooled to 2 GK, and the hadrons annihilated with the anti-hadrons, leaving a few hadrons. After this point, all the quarks had bound into hadrons.

At 1 second, the temperature was 10^{10} K, cool enough for the quarks to bind and form stable particles like protons and neutrons, and for the anti-quarks to form anti-protons and anti-neutrons. This caused the reaction in Equation 4.2 to go only one way, $q + \overline{q} \rightarrow \gamma + \gamma$, leaving the universe filled with bound quark states like protons and neutrons.

At the same time, the neutrinos', anti-neutrinos', and photons' energies became low enough for their mass to prevent photon or neutrino annihilation to electrons and positrons, and the reactions $e^+ + e^- \to \nu + \overline{\nu}$ and $e^+ + e^- \to \gamma + \gamma$ went entirely to the right, creating copious neutrinos and photons and leaving relatively few electrons and even fewer positrons. The universe was still filled with charged particles—mostly protons, electrons, and their anti-particles—and the photons remained trapped: Scattering off the charged particles kept the photons from traveling long distances from their point of origin. Once most of the electrons and positrons had annihilated to create photons and neutrinos, the neutrinos could only interact weakly with the protons and neutrons. The weak interaction did not take place fast enough to maintain thermal equilibrium between the neutrinos and other particles. The neutrinos and anti-neutrinos only interacted weakly and, unable to lose energy to other particles, fell out of equilibrium with the other matter in the universe. The neutrinos' temperature continued to decrease because of the expansion of the universe, and they went streaming off, forever on straight trajectories without colliding with other particles, much as the photons would when released by the formation of hydrogen 377,000 years later. The neutrinos and photons are both examples of particles "decoupling" from the rest of the matter in the universe: one kind of particle ceasing to interact much with all other particles and no longer playing a role in the evolution of the universe.

At the same time, the protons, anti-protons, neutrons, and anti-neutrons annihilated to form lighter particles. With both the leptons and baryons, a curious thing happened: For some reason, it turns out there were a "few" more protons than anti-protons and a "few" more electrons than positrons. How many? In the early part of the Big Bang, when the particle masses were small compared to their energies and did not influence the reaction rates much, there were roughly equal numbers of electrons, positrons, and photons. In the next 10 seconds of the universe's evolution, the expansion and cooling of the universe caused the electrons and positrons to annihilate to photons, but the reverse process—two photons forming an electron-positron pair—could not happen. The photons did not have the energy necessary to create the rest masses of the electron-positron pair. Electron-positron annihilation into two photons can always happen because the resulting photons have no mass. In those 10 seconds, all the positrons and almost all the electrons annihilated into photons, leaving about one extra electron per billion electron-positron annihilations.

The reason for the slight electron excess is unknown and is a great mystery of particle physics. We live in a world of protons and electrons that we call matter. The antiprotons and positrons, which we see only in high-energy **cosmic rays** and accelerator experiments, we call **anti-matter**. All the anti-matter we observe now came from the high-energy particles interacting with each other, producing particle–anti-particle pairs. For some unknown reason, there was one part per billion more protons than antiprotons before the era of annihilation between particles and anti-particles. Similarly, neutrons were left over from neutron–anti-neutron annihilation at the same level. The neutron is 0.14% heavier than the proton, and this mass difference caused the ratio of seven protons to one neutron 10 seconds after the Big Bang.

At 10 seconds, the universe had cooled to $T = 3 \times 10^9$ K, and most of the energy in the universe was in the form of photons. Since charged particles scatter light, the photons were rattling around, bouncing off the charged electrons and protons. This temperature was low enough to allow a proton and a neutron to fuse to make a bound state, in the process emitting an energetic photon: $p + n \rightarrow d + \gamma$. The **deuteron**, a bound state of a proton and a neutron, acts chemically just like hydrogen but is roughly twice as heavy. The deuteron has a very small binding energy, which means the strong interaction just barely holds the neutron and proton together. At this time in the universe, the deuterons were easily broken apart by interactions with other particles, so not many formed until the temperature dropped to about $T = 10^8$ K, about 3 minutes after the Big Bang. The deuterons then combined to form ^{4}He via $d + d \rightarrow {}^4\text{He} + \gamma$.

Seventeen minutes after the Big Bang, the universe had cooled enough to stop the production of ^{4}He. By this time, most of the remaining free neutrons, which have a lifetime of 15 minutes, had decayed. The universe was 75% hydrogen and 25% helium, with traces of beryllium and lithium from the fusion of ^{4}He. Of the hydrogen, 0.002% was deuterium. There was also 0.001% of ^{3}He and a tiny amount (0.00000001%) of ^{7}Li. The production of the elements in the early universe is called **Big Bang nucleosynthesis.**[15]

---

15. The creation of elements from protons also happens in stars and is called stellar nucleosynthesis. The theory behind stellar nucleosynthesis developed at about the same time as the theory of Big Bang nucleosynthesis. An important clue that Big Bang nucleosynthesis was needed came from the observation that

The Big Bang theory relates the ratios of D, ^{3,4}He, and ^{7}Li to hydrogen to the ratio of neutrons to protons, n_b/n_γ, in the early universe. Astronomers carry out measurements of D/H, ^{3,4}He/H, and ^{7}Li/H today, observing either objects very far away, and hence very old, or nearby objects unchanged since the Big Bang. In both cases, the measurement makes use of the absorption of the light produced in atomic transitions in the H, D, ^{3,4}He, or ^{7}Li in the surrounding gas. D/H gives the most precise result, since D's low binding energy precludes it from being produced in stellar nucleosynthesis, and the measured D/H ratio gives a proton/photon ratio in the range of 0.51 to 0.65 parts per billion. The measurements of CMB tell us the density of photons at the time neutral hydrogen formed, and we can work out that there were 10^{80} protons produced in the Big Bang in our presently visible universe. This comes to an average proton density of 0.28 protons per cubic meter. From measurements of distant supernovas, the microwave background, and the distribution of distances between galaxies described in Chapter 2, the total matter-energy density is equivalent to about 6 protons per cubic meter, so baryonic matter contributes about 4.7% of the energy content of the universe. Additionally, the measurements described in Chapter 2 showed that the total of dark and baryonic matter is 27% of the energy content of the universe. The bottom line: Not enough baryonic matter came out of the Big Bang to account for all the visible and dark matter observed in galaxies and clusters of galaxies, so

---

stars alone could not make enough ^{4}He to account for the amount observed in galaxies.

what we call dark matter cannot be failed stars, planets, or anything else made of baryons.[16]

4.2 NEUTRINOS AS DARK MATTER

In the 1980s, the idea was proposed that neutrinos could solve the dark matter problem: Although the neutrino's mass was known to be less than about 100 billionths of the proton's mass, the Big Bang model predicts that there would be about 300 neutrinos per cubic centimeter today. If the neutrino had a mass of even one-thousandth that of the proton, it would be enough to explain dark matter. This easy solution ran into two problems: Experiments measuring the neutrinos' mass showed that they were too light to be dark matter (or even a significant fraction of dark matter); and the low-mass dark matter particles would not have allowed galaxies to form, as we will see below.

The idea of a particle called the neutrino came up oddly. In the early 1920s, nuclear physicists observed beta decay in which a neutron was converted to a proton and emitted a charged particle called a **beta ray** (now known as an electron). **Alpha decay**, the emission of a helium nucleus in the decay of a radioactive nucleus, had been observed by Henri Becquerel in 1896, and the emitted **alpha particles** were known to be quite heavy—several proton masses. Becquerel also observed the emission of

---

16. Until recently, astronomers did not even know where all the protons were. Counting all the stars in the universe from galaxy surveys and including interstellar gas in galaxies accounted for about half the number of baryons that should have been produced by the Big Bang, resulting in the "missing baryon problem." Recent observations indicate that the other half of the baryons may be composed of hot gas between the galaxies.

much lighter particles. Around 1900, Ernest Rutherford termed emissions of heavy particles "alpha" radiation and the lighter particles "beta" radiation.

From the conservation of energy, the kinetic energy of an electron emitted in beta decay should always be the same energy, and it should be equal to roughly the mass difference of the initial and final nuclei minus the rest mass of the electron. However, this was not what was observed: In 1911, Lise Meitner and Otto Hahn showed that beta electrons were emitted over a broad range of energies rather than at a single energy. At the time, this problem created such a stir that several physicists proposed that beta decay violated energy conservation.[17] Wolfgang Pauli's solution to the problem in beta decay included the emission of a light neutral particle that he originally named the "neutron." James Chadwick's discovery in 1932 of a neutral particle of approximately the proton's mass caused a renaming: Chadwick's particle became the "neutron," meaning the "neutral one," and Pauli's the "neutrino," meaning "the little neutral one."

Subsequent experiments showed that there are three kinds of neutrinos, each associated with one of the charged versions of the electron, the muon, and the tau: an electron

---

17. This idea, by Niels Bohr, is not as farfetched as it seems. In the first and second decades of the twentieth century, the Theory of Relativity, which had overturned Newton's ideas about the absolute nature of space and time, was just gaining acceptance. Quantum mechanics, which undid many ideas about what could and could not be observed, was just being developed. This was a revolutionary time, and everything was being questioned. Bohr proposed that in beta decay and other quantum mechanical systems, energy conservation might only be statistically true and that any given decay did not need to conserve energy. In the mid-1920s, then, Charles Ellis showed that the beta spectrum had an upper limit, indicating that Bohr's conjecture was incorrect.

neutrino, a muon neutrino, and a tau neutrino. By the late 1990s, experiments showed that the neutrino mass must be very small—between a billionth and a trillionth of a proton mass. With about 660 neutrinos and antineutrinos per cubic centimeter created in the Big Bang, neutrinos do not provide enough mass to be dark matter. In any case, neutrinos cannot be responsible for the observed effects of dark matter, since they move too fast to remain gravitationally bound to the potential wells created by the quantum mechanical density fluctuations early in the Big Bang. The neutrinos from a region of higher density would have flown out of that region, reducing its density and lowering the gravitational potential, allowing even more particles to escape. This would eventually have had the effect of erasing the density fluctuations, and the universe would have become homogeneous, with the same mass density everywhere. What we observe is very different, with clusters and superclusters of galaxies of varying sizes. One could hypothesize that there could have been a fourth heavy neutrino, but observations at accelerators and nuclear recoil experiments in the early 1990s showed there were only three kinds of neutrino.

4.3 BLACK HOLES, WHITE DWARFS, FAILED STARS, AND PLANETS

4.3.1 Baryonic Compact Objects

Although the Big Bang theory was well developed by the early 1980s, measurement of the deuterium abundance in the universe was not accurate enough to determine whether the Big Bang could have produced enough baryons to account for the effects of dark matter. Many people at the time (like me) thought that dark matter could just be dark baryons: objects that do not produce light, like failed stars, planets, or neutron stars—baryonic compact objects.

Gravitational lensing provided a way to search for dense, compact objects. Collectively, these objects are known as **Massive Compact Halo Objects (MACHOs)**. They could be as light as 1/100 of a solar mass or as heavy as 100 solar masses.

Every once in a while, a MACHO will pass between an observer on Earth and a star in a small galaxy outside the Milky Way. If the MACHO passes close to the line of sight between the observer and the star, gravitational lensing will occur. The gravity of the MACHO focuses the star's light, making it appear perhaps hundreds of times brighter (Fig. 4.1). An experiment, also called MACHO, was able to detect brightness changes by 30%, the brightness change when a MACHO crosses almost exactly in front of a star. In the 1990s, the collaboration of astronomers working on the MACHO experiment set about to use an automatic telescope and observe enough stars each night to see whether MACHOs passed in front of them. The question was: "How many stars do we need to look at?"

The Milky Way has a total mass of 10^{12} solar masses, and if MACHOs with a mass of 0.1 of a solar mass make up part of the dark matter, then there are 10^{13} MACHOs orbiting around the galactic center, with a typical velocity of 200 km/s. If we look at a star outside the Milky Way in the LMC 150 klt-yr away, the Milky Way extends along 30 klt-yr of the line of sight. For these distances, a MACHO 12 klt-yr away from Earth in the Milky Way will have to

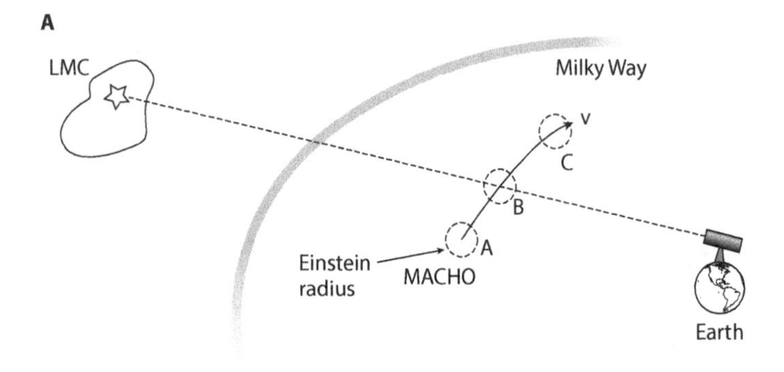

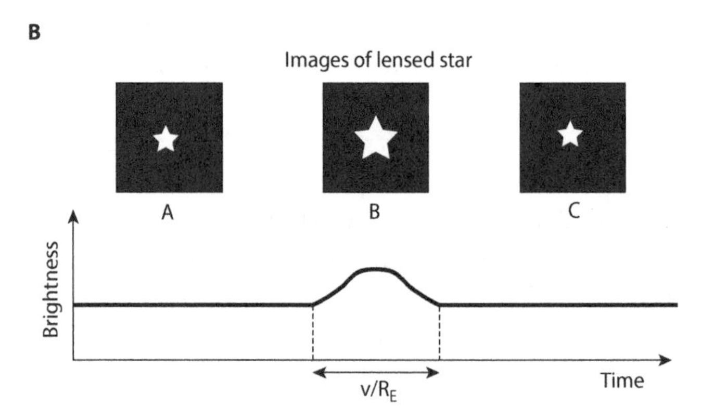

Figure 4.1. Lensing transiting objects in the Milky Way. A) An observer on Earth viewing a star in the LMC would see the star magnified by up to 30% if a gravitating object were to pass within one Einstein radius of the line of sight connecting the star to the observer. B) The change in the brightness of the distant star as the lensing object passes with time. For typical values of the masses and distances, the transit time would be several days.

pass within 6×10^{-6} lt-yr of the line of sight to be lensed from the Earth to the star. For a compact object 12 klt-yr from Earth and with the mass of our sun passing in front of a star 150 klt-yr from Earth, the Einstein radius is 6×10^{-6} lt-yr. The Einstein radius is how close the compact object has to pass to the line connecting the Earth and the distant star to be lensed. Working out the numbers gives 2×10^{-5} lensed MACHOs per year. Just staring at a single distant star will not work!

The solution is to look at a large collection of stars like the LMC. The MACHO collaboration made use of a 1.4-meter telescope in Australia to take repeated electronic images of the LMC each night over nearly 6 years. They sampled a total of 33 million stars, of which 18 million were analyzed in detail. For a given star, a plot giving the brightness of the star as a function of time was made over 5 years of observing. A star magnified by a passing MACHO would get up to 30% brighter when the Einstein radius of the MACHO crossed the line of sight (see Fig. 4.1). A MACHO moving at 200 km/s would take about 3 days to travel one Einstein radius, so the MACHO collaboration scanned their light curves for transit times ranging from 1 to 30 days.

MACHO and later similar experiments—Expérience pour la Recherche d'Objets Sombres (EROS, or "Experiment to Find Dark Objects"), Optical Gravitational Lensing Experiment (OGLE), Andromeda Gravitational Amplification Pixel Experiment (AGAPE), and others—observed many new kinds of variable stars (stars whose brightness varies with time, owing to their internal dynamics). These new variable stars made data analysis difficult.

Also, about half of all stars are part of binary systems consisting of two objects orbiting each other, causing a variation in the observed brightness of the system. Both binary objects and variable stars had to be identified and filtered out. With about 18 million stars to look at over 6 years, and a probability of 2×10^{-5} transits per year per star, MACHO experiment collaborators expected several dozen transits if dark matter were composed of MACHOs. Early on, the MACHO survey observed about a dozen events whose transit times indicated that the lensing objects had a mass of around half a solar mass. This transit rate would correspond to MACHOs accounting for about 20% of the galaxy's dark matter. However, subsequent measurements by MACHO, EROS, and OGLE experiments found far fewer transits, leading to the conclusion that dark matter cannot be massive baryonic objects the size of our sun. The first measurement was a statistical fluke.

4.3.2 Primordial Black Holes

The transient searches for compact objects, including black holes, ruled out black holes with masses above 0.1 solar masses as dark matter. Since the Big Bang did not make enough baryons to account for dark matter, and since black holes come from supernovas that form from baryons, it would seem fruitless to look to black holes as a solution to the dark matter problem. However, PBHs that formed in the primordial universe *before* the baryons may be an exception.

In 1974, Hawking showed that black holes could have formed from quantum mechanical density fluctuations of whatever composed the universe before baryons were created. Black holes form when there is enough mass inside a sphere of a certain radius that the gravitational field at the surface of the sphere is strong enough to bend the light emitted from the sphere back inside the sphere. Figure 4.2 shows how the apportionment of energy changes during the evolution of the universe depending on whether dark matter is particles (WIMPs or axions, discussed in detail in Chapters 5–7) or PBHs. The universe may have diverted energy into dark matter in the form of PBHs before the creation of baryons. In recent years, some theorists have calculated how PBHs might have formed in the early universe, but describing this era of the universe needs both quantum mechanics and gravity, and we do not have a theory of quantum gravity. Thus, the models that theorists create contain many assumptions about how gravity and quantum mechanical particles work together. However, just because we cannot calculate how PBHs may have formed does not mean they cannot exist.

How PBHs might manifest themselves depends a great deal on their mass. If enough PBHs were created in the early universe before the creation of normal matter, then the heavier each PBH is, the fewer there are of them. The lightest allowed PBHs that do not evaporate in the age of the universe would have a mass of 10^{15}g and a diameter of less than a millimeter. One would hit the Earth every million years. Not much would happen: Moving at one-thousandth the speed of light, a 10^{15}g PBH would cross the Earth in about 42 seconds if it went from the North Pole to the South Pole, and it would sweep up less than a milligram of material. The heaviest allowed PBH, 10^{26}g, about

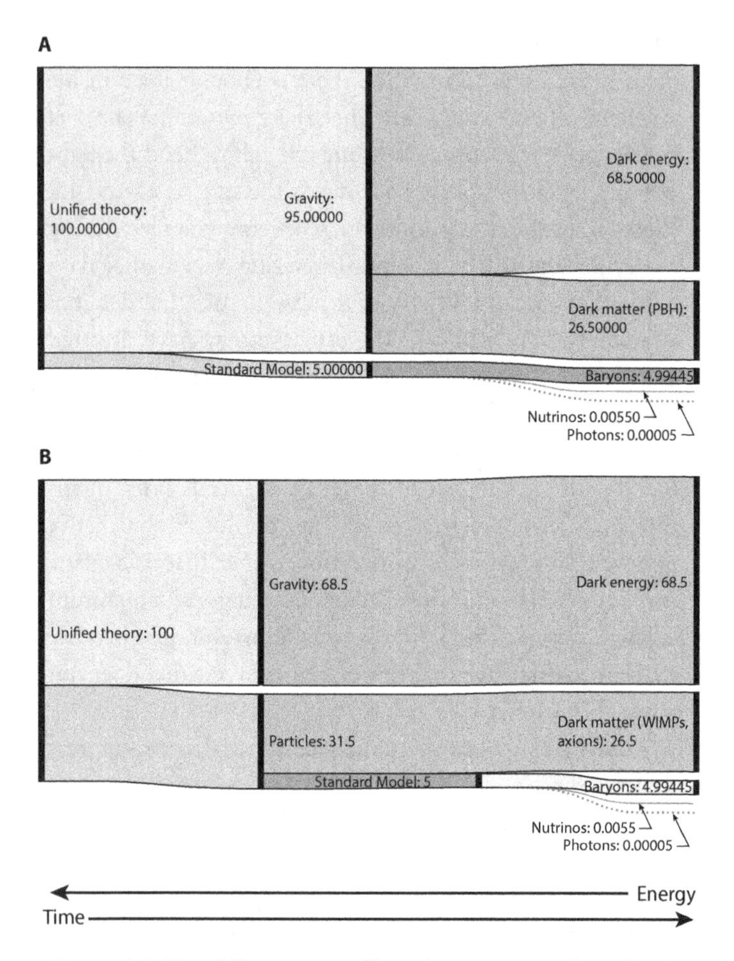

Figure 4.2. Two different views of how the universe might evolve. In both cases, a unified theory describes the universe in early times (the left side of the figure). As time evolves, the unified theory divides into a theory of particles and a theory of gravity. A) If PBHs are dark matter, then they are described by gravity, and the Standard Model does not need to be changed. B) If dark matter consists of particles, something must be added to the Standard Model to give a particle theory at early times in the universe.

the mass of Mercury, would collect a trillion kilograms of the Earth as it passed through. Heavier PBHs would have been observed by transient lensing surveys. However, such a heavy PBH has less than a one-in-a-million chance of hitting the Earth in the age of the universe.

What could transient surveys say about PBHs? Since observations imply that PBHs have masses that are a trillionth to a billionth of the masses of black holes from supernovas, the lensing of a star by a PBH occurs over minutes or hours, rather than the several days of the solar mass–sized black hole. The much smaller mass of a PBH makes the brightening of a star smaller and less frequent. A transient lensing survey would require viewing 100,000 times as many stars over a shorter time with a more sensitive camera than used in the MACHO experiment. Repeat observations would need to be made every few minutes. A team working at the Subaru telescope on the island of Hawai'i has begun just such a survey by looking for transient signals in stars in the galaxy Andromeda, 2.5 Mlt-yr away. The Subaru team measured the light from 87 million stars every 2 minutes for 7 hours. If PBHs were dark matter, the team should have observed about 1,000 transits. Instead, in the first results, Subaru observed one, showing that, when combined with other measurements, if PBHs were dark matter, their mass would have to be less than the mass of an asteroid 100 km across. The Subaru team is continuing to analyze the data from their survey.

PBHs still could be dark matter, but they would have to have masses between 10^{15}g and 10^{19}g, corresponding to the masses of asteroids 4 km and 100 km across, respectively. Hawking predicted that PBHs with masses less than

10^{15} g would radiate away. Transits in this mass range would be hard to detect, as the transit times would be seconds or less and the brightening of the background star just a few percent at most.

In many ways, PBHs are perfect dark matter candidates: We have no way of detecting them except by their gravitational influence on normal matter, described in Chapter 2, and they are formed early enough so we do not need to add anything to the Standard Model to explain dark matter. PBHs shift the problem of dark matter to the problem of unifying quantum theory with gravity by shifting the dark matter from a particle to a feature of gravity, along with dark energy (Fig. 4.2).

4.4 MODIFIED NEWTONIAN DYNAMICS

Rather than introducing a new particle, Mordechai Milgrom in 1981 suggested changing Newton's second law, $\vec{F} = m\vec{a}$. Newton's second law works very well on Earth and in our solar system, so any change would have to be important on a distance scale of a galaxy or more. Milgrom's approach is to keep the second law for large accelerations and modify it for smaller accelerations, so that the same force produces less acceleration for galactic-scale distances of 100 klt-yr. The change in acceleration at these distances could explain the observations of the galactic rotation curves described in Chapter 2.

However, this modification does not explain the evidence for dark matter on the scale of a galaxy cluster or larger; this would require an even smaller value for long-distance acceleration. Also, the observation of 1E 0657-56 used the x-rays emitted by the hot baryonic matter to show that baryonic matter was in the wrong place to cause the gravitational lensing of light passing through the colliding clusters. The lensing matter and baryonic matter had become displaced by the collision, and no sensible modification of Newton's laws could explain that. The Big Bang theory gives a consistent picture of the creation of all baryonic particles in our universe, and many measurements have verified the theory.

From the original observation of dark matter in the 1930s to the mid-1990s, all known forms of matter were eliminated as explanations for the dark matter problem. At the same time, modifications to Newton's second law proved untenable. Where does that leave us? The conclusion is that our understanding of particle physics is incomplete, and we will need to add something to the Standard Model to explain dark matter. Thus we must look for supersymmetric particles, axions, or the dark force carrier. Many researchers have undertaken this task that, so far, has proven daunting. We will next explore the recent efforts underway to answer the question: What is dark matter?

5
SEARCHING FOR WIMPS ON EARTH

This chapter explores some of the experiments looking for dark matter in our galaxy. Astronomical observations have provided conclusive evidence of dark matter's existence, and experimenters now face the challenge of measuring its properties: how dark matter interacts; the mass of dark matter particles, if dark matter is made of particles; what its density and velocity in the galaxy are; and a host of other questions. Nature does not guarantee success: Dark matter could simply be something that *only* interacts through gravity, or its nongravitational interactions could be so small that they can never be detected experimentally. Our curiosity drives us to be optimistic and look where we can for answers.

The main dark matter candidates are axions, particles that result from the strong interaction, and heavy particles from extensions to the Standard Model. Before delving into the experiments, we need to understand a little more about what astronomers know about dark matter near the Earth. Sections 5.1 and 5.2 describe the way dark matter might weakly interact with normal matter and be detected in an experiment. Sections 5.3 and 5.4 describe several different kinds of dark matter experiments currently operating.

5.1 DARK MATTER IN GALAXIES

The hunt for dark matter starts with a simple picture of our galaxy, which includes both dark and baryonic matter. Chapter 1 developed the simplest picture of the Milky Way consistent with most observations. The baryons lie in a disk about 3 klt-yr thick, with a radius of about 50 klt-yr. The Earth orbits about 24 klt-yr away from the center of the galaxy, moving at 220 km/s and completing one orbit around the galactic center every 240 million years. A very dense region of stars and dust surrounds a 4 million solar mass black hole sitting at the center of the Milky Way, with a spherically shaped bulge extending out to about 1 klt-yr. The dark matter lies in a large, approximately spherical distribution, or "halo," extending out beyond 150 klt-yr, well beyond the visible edge of the baryonic matter. At the radius of the Earth's orbit, the dark matter density is roughly one-third to one-half of a proton mass per cubic centimeter. Researchers believe that the dark matter orbits the center of the galaxy more slowly than the baryons.

Here's why astronomers believe that the dark matter orbits more slowly than the normal matter: Galaxies began with baryonic gas and dark matter remaining from one of the quantum mechanical density fluctuations from the Big Bang. The gas and dark matter both fell inward toward the center of the overdense region but behaved differently. Gravity pulled the dark matter in toward the center of the overdense region, but the dark matter did not interact significantly with itself or baryonic matter, so the dark matter could not lose energy, fall toward the center of the galaxy, and remain there. The dark matter's angular momentum balanced the pull of gravity, so the dark matter could orbit around the center.

The baryons, however, could interact by scattering off one another via the strong or electromagnetic interaction, causing the baryons to lose energy in various ways, so that they could fall toward the center of the galaxy. A proton, for example, may have lost energy by hitting another proton and creating a neutral pion from the energy of the collision. The neutral pion decayed to two photons, which escaped and carried energy out of the galaxy. Or the proton might have knocked into another proton and sent it flying out of the galaxy, carrying some of the first proton's energy with it. A proton scattering a microwave photon to an x-ray photon that then escaped the galaxy is another way that baryons could lose energy. In each case, the proton lost some energy and fell toward the center of the galaxy into a smaller orbit.

Baryons also lose angular momentum through collisions, but much more slowly than they lose energy. As time passes—hundreds of millions of years—most of the baryon's energy resides in its angular momentum, which makes it follow a circular orbit. Eventually, most baryon orbits in a galaxy are nearly circular. Baryons keep most of their angular momentum and orbit faster as they fall closer to the center of the galaxy. The baryons end up in a flat disk rotating through the much larger, more slowly moving dark matter halo.[18] Some baryons form stars, others remain as interstellar gas.

---

18. The formation of galaxies is something that is still not well understood. There is clear evidence that mergers between galaxies play a major role. For example, the Milky Way swallowed a galaxy about 100 million years ago and will merge with Andromeda in about 5 billion years.

The complete picture of the Milky Way is probably much more complex. Computer simulations predict that there should be many small satellite halos of dark matter surrounding the Milky Way. Other simulations predict very high-density waves of dark matter called **caustics** propagating back and forth through the galaxy. Still other models predict that there should be clumps of dark matter, remnants of galaxies captured and torn apart by the Milky Way. The distribution of dark matter in our galaxy is not known in detail; observations of the motion of stars near the Sun give a dark matter density of about one-half of a proton mass per cubic centimeter, with an uncertainty of about 10%. The balance between the pull of gravity on dark matter and the dark matter angular momentum means that the typical dark matter velocity is around 1/1000 the speed of light. The dark matter moves randomly in all directions, like gas in a bottle, and the baryons orbit with an average velocity of 220 km/s, relative to the dark matter. We cannot say much more with certainty until we understand what dark matter is.

5.2 DETECTING WIMP DARK MATTER FROM ELASTIC SCATTERING

Armed with some understanding of the velocity and density of dark matter near Earth, the mechanism for detecting dark matter presents the next challenge. Direct detection of axions, WIMPs, or other dark matter particles means measuring the energy that has been transferred from a dark matter particle to a normal matter particle, referred to as the **recoil particle**. A dark matter search experiment consists of a mass of normal matter particles for the dark matter particles to interact with, referred to as the "target," and some means of measuring, or "reading out," the recoil particle's energy after an interaction with dark matter.

The typical kinetic energy that a nucleus would have after being struck by a WIMP is about the same as an x-ray photon: a few thousand times the energy of a visible photon. Ensuring that as few as possible normal matter particles transfer energy to the target particles poses a major challenge for experimenters. A state-of-the-art dark matter experiment will discriminate against normal matter interactions, so that the experiment will detect only a few normal matter interactions in each kilogram of target material in a year. Poor design or a careless choice of materials could leave the experiment swamped in recoils from normal matter interactions—alpha or beta decays, for example—and unable to detect dark matter–induced recoils. Many dark matter experiments take place several kilometers underground, so that the ground above the experiment absorbs the cosmic rays from space. Very pure materials—usually a combination of copper, lead, and water—form a "shield" against radioactivity from the walls, ceiling, and floor in the experimental area. The detector components inside the shield are also made from very pure materials that have been carefully tested and selected for low levels of radioactivity. The following text describes experiments on Earth that hunt for dark matter, focusing on WIMPs.

"WIMPs" refers to a general class of dark matter particles ranging in mass from a few to thousands of proton masses. WIMPs have no electrical charge and very small or no interaction strength with normal matter, aside from gravity, itself a feeble interaction. In the late 1980s, when dark matter experiments were getting started, many researchers believed that very massive neutrinos could be WIMPs. Early experiments showed that the interaction strength of neutrinos was much too large for a neutrino to be dark matter. The neutral supersymmetric particles described in Chapter 3 provide excellent WIMP candidates, and much of the theoretical dark matter work focuses on supersymmetric WIMPs.

Elastic scattering provides the simplest interaction between WIMPs and baryonic matter (Fig 5.1). A WIMP simply hits an atomic nucleus in the target, initially at rest, transferring some of its energy to the nucleus and bouncing off the nucleus. The atomic nucleus recoils, hitting the nearby atoms, losing some energy with each collision. Advances in detector technology have made measuring the small energy of a single recoiling nucleus possible by collecting a few hundred to a few thousand electrons, photons, or vibrations from the recoiling nucleus bouncing off the surrounding matter.

The number density of dark matter particles, number of target nuclei, and interaction probability determine dark matter's interaction rate with normal matter. To see how this works, imagine a parking lot with a few cars scattered around the parking spaces. You blindfold a friend, put them in a car, and let them loose to drive around without being able to see where they are going until they hit one of the parked cars, which will happen, on average, after some time period that depends on the number of parked cars, their size, the size of your friend's car, and so on. If you then double the number of parked cars, the

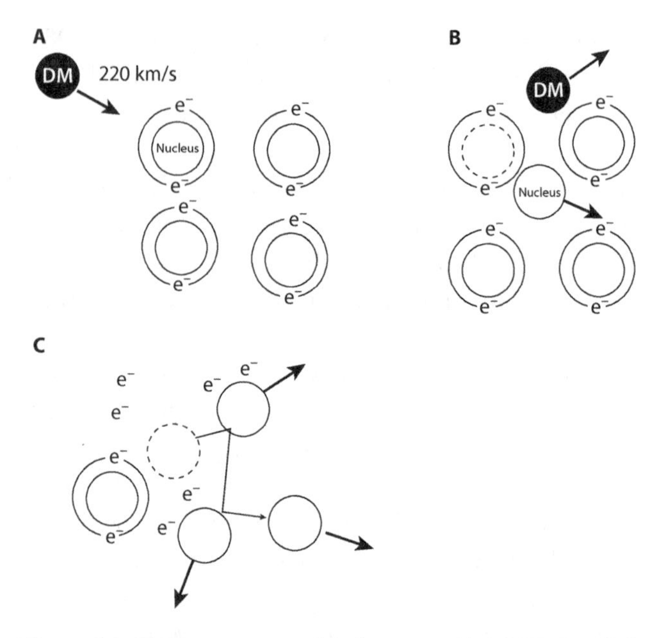

Figure 5.1. Elastic scattering of dark matter. A) A WIMP collides with a crystal of germanium, for example, at 220 km/s. B) The WIMP strikes a germanium nucleus, giving it a fraction of its energy and knocking off its electrons. C) The struck nucleus bounces off surrounding nuclei, knocking off their electrons and causing them to recoil. This process continues. About one-third of the energy of the struck nucleus goes into knocking the electrons off their atoms, and the other two-thirds goes into the motion of the nuclei, which heats the germanium crystal. A typical dark matter interaction will knock off up to a few thousand electrons.

average time to hit one will halve. If you keep the number of parked cars the same, but double the number of moving cars, the time until someone hits a car will again halve. From this thought experiment, we can see that the time for two cars to collide is proportional to the number of parked cars and, separately, proportional to the number of moving cars.

Dark matter fills the role of the moving cars, and the target particles are the parked cars—double either and the time to collision halves. Experimenters like to talk about rate, the reciprocal of the average time to a collision, and doubling either the number of target particles or the number of dark matter particles will double the collision rate.

Nature dictates the number of dark matter particles passing through the detector and the probability that they interact with atoms in the detector, but experimenters can control the number of target particles, so experimenters want to build large experiments to have as many target particles as possible.

Particle detectors made of germanium that detect electrons and photons are common in nuclear physics, and many companies know how to build them at a reasonable cost. (For my Ph.D. project, a company built a 1.8-liter germanium detector with 10 kg of target mass for a few hundred thousand dollars in 1986.) The detector must operate for a year in an underground site. Graduate students are relatively inexpensive to employ, and an appropriately supervised graduate student can operate the detector and analyze the data, so what's the big deal?

It may seem like a good idea, but backgrounds—known processes that mimic the nuclear recoils caused by dark matter—are like false positives in cancer screening: Their interactions look exactly like the dark matter recoils that the experiment is supposed to detect, but they come from some mundane source, like cosmic rays or low levels of radioactive contamination. **Gamma rays** are far more common than dark matter, and they are much more likely to interact in the detector. Gamma rays are very energetic photons emitted in nuclear transitions, usually following radioactive decay. In a typical room, a few hundred gamma rays go through one liter of germanium each second, and they have a probability of interaction of at least a few percent. The gamma rays come from all around, since radioactive elements contaminate most materials at a low level.

It gets harder: Cosmic ray particles also pass through the detector, and they have a 100% chance of making a signal. Cosmic rays are charged particles that are accelerated in supernova explosions and travel throughout the galaxy. Some of them hit the Earth, pass through the atmosphere, and reach the ground. About 100 cosmic rays pass through a square meter of the Earth's surface every second, which translates to about two cosmic rays per second passing through the liter of germanium.

Suddenly, looking for dark matter has gotten very difficult: In a year, there might optimistically be 20 or so dark matter interactions, 30 million cosmic rays, and even more interactions from gamma rays from low-level contamination of radioactive materials. Most radioactive materials also emit **alpha rays** and beta rays that add to the backgrounds by mimicking a dark matter signal.

The first dark matter experiments, like my Ph.D. project in the late 1980s, were designed to look for a rare process called **double beta decay**. Double beta decay experiments face similar obstacles as those faced in dark matter experiments. One can escape cosmic rays by doing the experiment underground—for example, in a road tunnel in Switzerland underneath a 1 km mountain. The rock of the mountain absorbs the cosmic rays and reduces the number of interactions to a few dozen per year at energies higher than the dark matter interaction energy. However, gamma rays pose a more difficult problem. First, the detector itself must be constructed of materials with the smallest amount of radioactivity, using high-purity copper, Delrin plastic, and germanium. Experimenters building "WIMP detectors," or double beta-decay experiments, test all the materials needed to build the experiment before using them. Choosing the wrong material might swamp the experiment in background recoils.

That takes care of the gamma rays inside the detector, but what about gamma rays coming from the walls and air around the detector? For these, high-purity lead will absorb 99% of the gamma rays. However, lead itself emits many low-energy gamma rays, so most experiments need a few inches of high-purity copper inside the lead to protect the target material from the lead itself. Figure 5.2 shows a typical experimental arrangement.

Several other tricks can be used, such as using positive gas pressure to keep radioactive radon gas from seeping into the detector,[19] microphones to identify small vibrations, and scintillator particle detectors surrounding the outer lead shielding to signal the passage of rare cosmic rays.

---

19. Uranium and thorium contaminate most rock and concrete and decay to radon, a radioactive gas. The radon seeps out of the concrete and can diffuse through the lead and copper shield to get close enough to the detector to fake a signal.

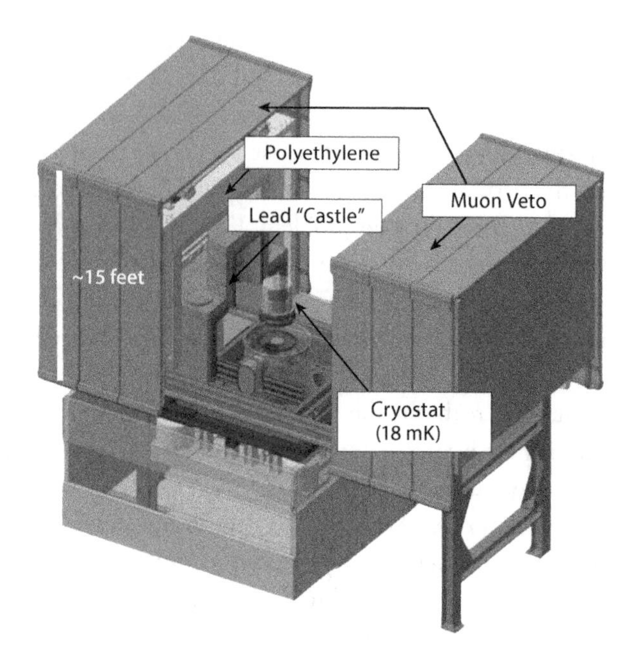

Figure 5.2. Schematic of an underground dark matter experiment. The dark matter detector (labeled Cryostat) is located inside approximately 30 cm of copper that is, in turn, surrounded by up to 50 cm of lead (labeled Lead "Castle"), followed by a layer of neutron absorber (labeled Polyethylene). The **muon veto** surrounds the entire apparatus and signals the passage of cosmic ray muons. Dark matter detectors also frequently have a 1- to 2-meter water shield that acts as both a neutron absorber and muon veto. The dark matter detector is below a few kilometers of rock to absorb cosmic rays. The polyethylene acts to absorb neutrons from radioactivity in the rock and cosmic ray interactions. The lead absorbs gamma ray photons from the surrounding rock, and the copper absorbs gamma ray photons from the lead and internal contaminants.
Source: Edelweiss-III Collaboration.

5.3 MEASURING TWO KINDS OF ENERGY

A very important technique has emerged over the past 20 years, stemming from the need to further reduce false signals from alpha, beta, and gamma rays. Even with the best materials and thickest shields, some residual radiation that emits gamma rays will enter the detector and fake a dark matter signal. The original germanium experiments simply measured the energy deposited in the germanium and did not make use of a key element: Gamma rays produce a signal by interacting with the atomic electrons, while the dark matter interacts by hitting the atomic nucleus. Let us look at these two processes more carefully.

A gamma ray has a large probability of hitting an atomic electron in the detector, giving the electron some energy. The recoiling electron then travels through the detector material, knocking electrons off other atoms. The electrons are much lighter than the surrounding germanium nuclei, so not much energy is transferred to a nucleus when an energetic electron hits it. Thus an energetic electron does not make vibrations in the germanium the way that a recoiling nucleus does. The detector then collects these electrons and, by counting them electronically, determines the energy of the gamma ray. In contrast, a dark matter particle will most likely hit a nucleus, which then recoils and bounces off other nuclei, knocking electrons off the atoms and making vibrations, called **phonons**, in the target material (Fig. 5.3). For a gamma ray, all of its energy goes into knocking electrons off atoms; while for a recoiling nucleus, about one-third of the energy goes into

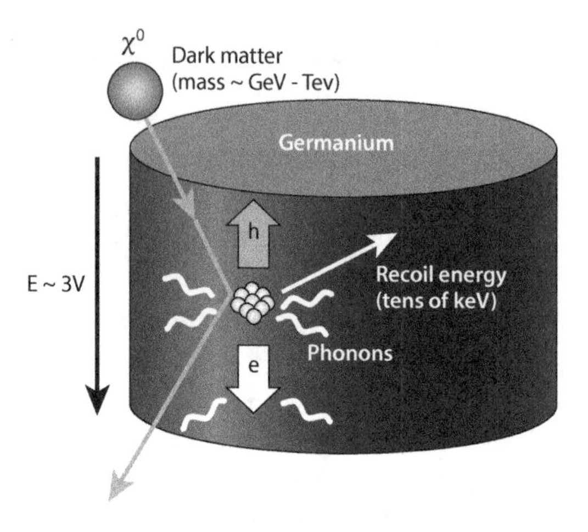

Figure 5.3. Principle of two energy measurements pioneered by the Cryogenic Dark Matter Search (CDMS) and used to discriminate between signal and background interactions. Second-generation dark matter detectors make use of the fact that dark matter particles cause the nuclei to recoil, while gamma ray photons cause atomic electrons to recoil. A recoiling nucleus leaves a very different signature in the detector than that of a recoiling electron. There are several schemes for measuring the energy deposited by the recoiling nucleus two different ways, and the figure shows an early method that measured the number of electrons (e) and holes (h), and the heat generated by interactions in a germanium crystal. The electrons are pushed to the top of the crystal by an applied electric field and collected there. Highly sensitive thermometers on the bottom of the detector make a signal if the heat from the recoiling nucleus reaches them. For a dark matter interaction, one-third of the energy goes into the electrons, and two-thirds goes into heat. For a photon interaction, all the energy goes into the electrons.
Source: From "Cryogenic Dark Matter Search (CDMS)" (Standard University for the U.S. Department of Energy Office of Science).

electrons and two-thirds goes into phonons in the target material.[20]

Measuring both electron and phonon energy provides a powerful discriminant between dark matter interactions and gamma ray interactions. The CDMS and Cryogenic Rare Event Search with Superconducting Thermometers (CRESST) experiments, as well as several others, pioneered the technique (now considered an essential feature of any dark matter experiment) in the early 1990s. Counting the phonons requires that the target mass be kept a few hundredths of a degree above absolute zero. Superconducting sensors cover the detector to pick up the vibrations from the struck nucleus careening off other nuclei. Figure 5.3 shows a schematic of the CDMS detection principle.

Even using the two energy measurement techniques, the detector had to operate at the bottom of a deep mine to avoid being swamped by background from cosmic rays. The first generation of detectors weighed a few kilograms—about the same as the original germanium double beta decay experiments—but had 10 million times better sensitivity to dark matter because of the ability to discriminate between nuclear recoils and other types of interactions. Nuclear recoils from dark matter have not been conclusively observed, and the current experiments are sensitive to interaction strengths about 10^{-9} that of a neutrino, the most weakly interacting known particle. Neutrinos from an accelerator can pass through a distance equal to 200 Earth diameters without hitting a nucleus. WIMPs, if they exist,

---

20. The fraction of energy going into electrons for dark matter–type interactions varies considerably with energy.

could pass through two trillion Earth diameters without hitting a nucleus.

Improving the sensitivity requires larger detectors, to give the dark matter more target nuclei to hit. In addition to the solid-state detectors like CDMS, current-generation WIMP detectors use relatively inexpensive liquid noble gases, such as xenon, argon, or neon, as a WIMP target. A cooling system keeps the noble gas a liquid. Instead of collecting the electrons, some of the experiments make use of the light emitted when the electrons are torn off the atoms, known as **scintillation light**. Light detectors detect and count the photons and are cheaper and easier to operate than the superconducting thermometers of the earlier generation of WIMP detectors. The principle of two energy measurements to reduce the gamma backgrounds remains. In noble gases, a recoiling nucleus creates visible light, in addition to knocking out electrons. In these detectors, the ratio of electrons to photons identifies nuclear recoils. An interesting aspect of these detectors is their size: In a detector with 100 kilograms or more of target material, which measures about 2 m across, the noble gas itself acts as a shield against neutrons and gamma ray photons. The electronic or light signals are used to triangulate the position of an interaction, and only those interactions away from the edge of the target region are treated as candidates for dark matter interactions (Fig. 5.4). Current experiments use several hundred kilograms of noble gas but select interactions from the central target region, using the outer target region as shielding.

Ultimately, two things limit the nuclear recoil technique for searching for dark matter. First, even deep underground,

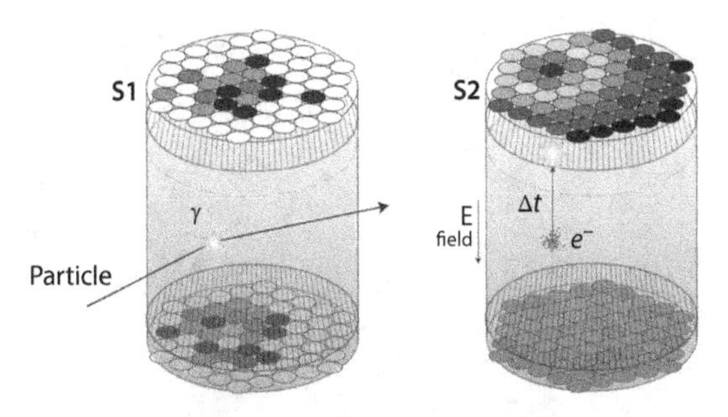

Figure 5.4. Triangulation of signals to locate the dark matter interaction. The arrival time of signals from an interaction in the detector may be used to locate the dark matter interaction position inside the detector. In the case shown, a dark matter interaction causes the emission of light in a liquid noble gas. The light is detected by photon detectors located above and below the liquid. The photon detectors can record the arrival time with an accuracy of 0.1 billionth of a second, corresponding to a distance of 3 cm. Light detectors located above and below the liquid volume collect the light generated as the recoil particle stops, referred to as the "S1 signal." An electric field pushes the electrons liberated by the recoiling particle to an amplification region at the top of the volume, where light is emitted and detected, forming a second signal called "S2." Comparing the two signals separates dark matter–induced recoils from recoils caused by other particles, such as gamma rays.
Source: David Malling and Carlos Faham. CC BY 3.0 by Gigaparsec.

there are a few neutrons from cosmic ray interactions in the rock surrounding the underground experimental area. Neutrons will interact in the detector just like dark matter, and they must be absorbed by external shielding. Water is an excellent neutron absorber, because when a neutron bounces off a proton in one of the hydrogen atoms, it gives up more of its energy than if it had hit a heavier

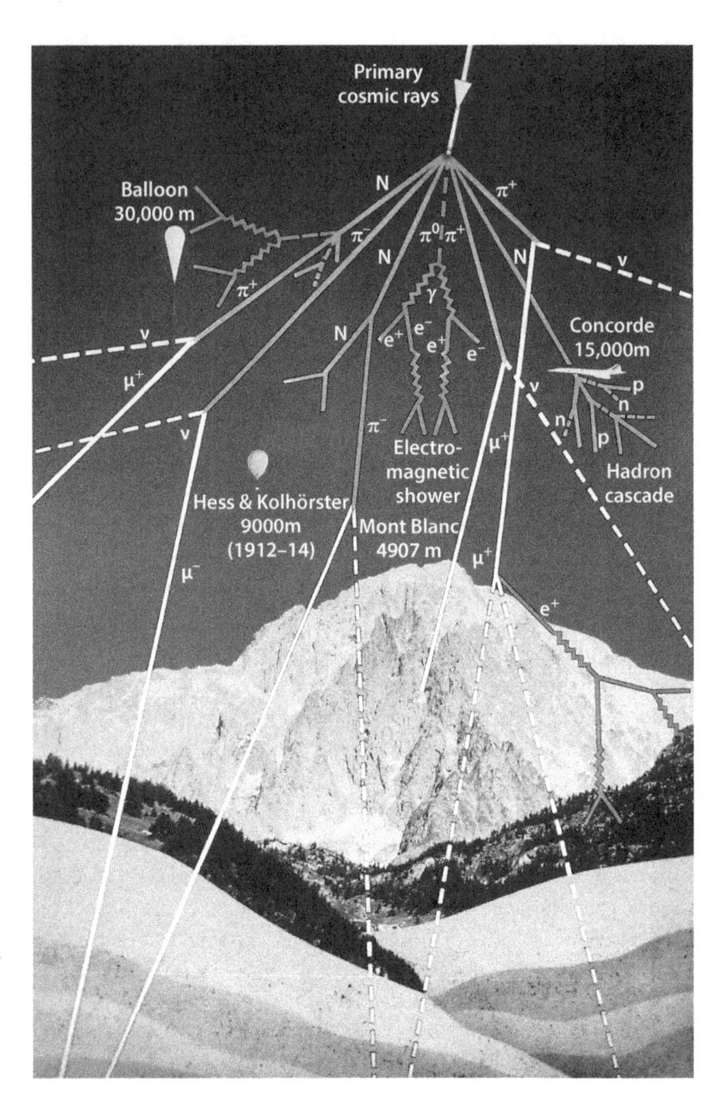

Figure 5.5. Neutrinos from a high-energy air shower. A cosmic ray proton moving at close to the speed of light strikes a nitrogen nucleus in the upper atmosphere, causing a cascade of particles. The unstable particles produced in the cascade quickly decay, emitting neutrinos, photons, protons, and other particles. Only the stable particles and muons reach the Earth. The first few dozen meters of the Earth's

nucleus. This is because protons and neutrons have about the same mass, and a neutron hitting a proton loses up to half its energy by transferring it to a proton. At lower energies, a proton and neutron fuse to produce a deuteron, $p+n \rightarrow d+\gamma$. Neutrons entering a thick water shield will hit protons in the water, losing energy with each collision; and, eventually, the neutron will fuse with a proton in the water. Shielding a few hundred kilograms of liquid noble gas with a few meters of water makes for a large apparatus but lies within the realm of possibility.

If the current generation of dark matter detectors does not observe dark matter, the next generation of WIMP detectors will need to have several tons of mass, requiring water shields of hundreds of tons. The nuclear recoil detectors and their large water shields must be built and operated deep underground. Constructing large underground caverns, which would have to span 20 or 30 meters, is now under way; and deployment of several tons of target material will take place in the mid-2020s.

Neutrinos from cosmic rays and the Sun present the second experimental limit. When a cosmic ray hits an atom in the Earth's atmosphere, the nucleus gets blasted apart, and a big spray of particles develops, producing some neutrinos that are referred to as **atmospheric neutrinos** (Fig. 5.5).

Figure 5.5 (continued). surface absorb the protons, electrons, and gamma rays. Muons penetrate several hundred meters and produce neutrons deep underground by interacting with atoms in the rock. Almost all the neutrinos pass through the Earth without stopping. Source: "High energy cosmic rays striking atoms at the top of the atmosphere give the rise to showers of particles striking the Earth's surface" © 1999–2021 CERN.

Neutrinos can travel through the Earth, so even an experiment deep underground will experience a rain of neutrinos. Enough neutrinos will interact in a several-ton detector to make a false signal, presenting a seemingly insurmountable problem: Recall that absorbing a neutrino requires several light-years of concrete. Dark matter experiments with target masses of a few tons will begin to face neutrino backgrounds in the coming years.

5.4 DETECTING THE EARTH'S MOTION THROUGH THE DARK MATTER HALO

The detectors described so far do not make use of an important feature of dark matter: In the simplest model of our galaxy, the baryons (including our solar system) move through the dark matter halo at an average velocity of about 220 km/s. If you look at the orbit around the galaxy that our solar system follows, the stars making up the constellation of Cygnus (the Swan) are ahead of us in the same orbit, so if we could "see" dark matter, it would look as though it was coming from Cygnus. As the Earth rotates around its axis, Cygnus traces out a circle on the sky each day, and the average direction of the incident dark matter particles would appear to follow Cygnus. Dark matter has a range of different velocities—from 0 to 500 km/s, with an average of 220 km/s—so not all of the dark matter would seem to come from Cygnus, but most dark matter particles would.

The Earth orbits the Sun with a speed of 30 km/s (Fig. 5.6A). Therefore, in November, when the Earth is moving away from Cygnus, the average dark matter particle speed falls to 190 km/s; and in May, when the Earth moves

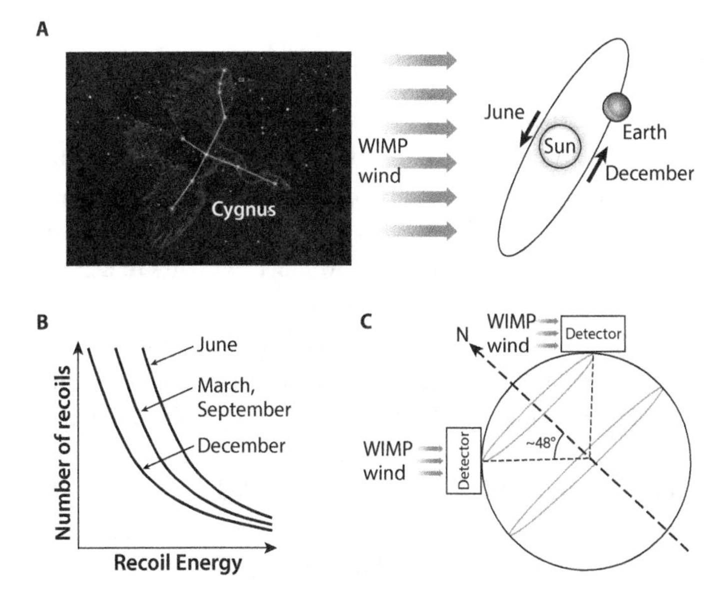

Figure 5.6. Two effects of the Earth's motion through the dark matter halo. A) The Earth moves at 30 km/s in its orbit around the Sun, and the solar system moves at 220 km/s through the galactic dark matter halo. In December, the Earth moves against the direction of the solar system at a relative velocity of 250 km/s, relative to the dark matter halo. Six months later, in June, the Earth moves along the solar system direction through the dark matter halo at a relative velocity of 190 km/s. B) In December, the relative velocity is higher, so the nuclei in the detector get hit harder, leaving more energy in the detector. In June, the relative velocity is lower, and less energy is left in the detector. All detectors measure total interactions above threshold energy. In December, the detector measures more interactions above threshold than in June. C) During a day, the Earth rotates on its axis, so a detector able to measure the direction of the recoiling nucleus would measure recoils coming from the side of the detector at one point in the day and from the top of the detector 12 hours later.
Source: The constellation Cygnus. NASA; A2L From "Annual Modulation of Dark Matter: A Review" by Katherine Freese, Mariangela Lisanti, and Christopher Savage (2013).

toward Cygnus, the average speed increases to 250 km/s. The kinetic energy of the dark matter particle depends on the square of the Earth's velocity through the dark matter halo, and the 10% variation in average speed measured on Earth between May and November translates to a 20% variation in WIMP kinetic energy.

The larger WIMP average kinetic energy in June leads to larger average recoil energy in May compared to November (Fig. 5.6B). Since a WIMP detector counts recoils with energies above a minimum energy, called the **energy threshold**, the number of recorded recoils will vary throughout the year, with the largest recoil rate happening in May and the lowest in November.

The Dark Matter/Large sodium Iodide Bulk for RAre processes (DAMA/LIBRA) experiment in Italy has published observations of the variation in count rate above the energy threshold over more than 11 years of continuous operation. The variation is consistent with a dark matter wind. The DAMA/LIBRA detector, made of sodium iodide crystals that emit light when a nuclear recoil occurs, resides in the Gran Sasso laboratory in Italy. Figure 5.7 shows the DAMA/LIBRA recoil rate over its period of operation. The recoil rate varies during the year, with the maximum recoil rate in June and the minimum in November, as would be expected if the recoils were caused by dark matter interactions.

Temperature or pressure variations and other effects may cause small changes in the energy threshold with time, and ensuring a stable energy threshold requires careful monitoring and periodic calibration. Experimenters work hard to achieve the lowest possible thresholds, getting right to the

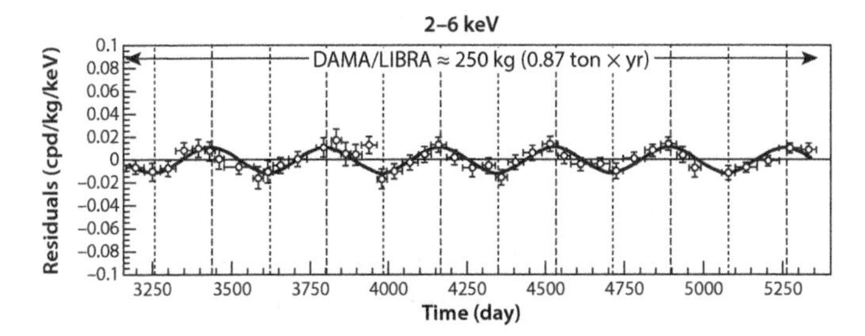

Figure 5.7. DAMA and DAMA/LIBRA data. The DAMA and DAMA/LIBRA experiments measured the total number of interactions above an energy threshold for more than 11 years. As described in Fig. 5.6, the number of interactions varies with the time of year. The count rate varies by a few percent over a year.
Source: arXiv:1002.1028v1.

limits of electronic technology and physical understanding. Certainly, some environmental parameters could have annual variations, leading to an annual variation in the measured recoil rate. For example, the average daily temperature varies over a year by several degrees, even in a location deep underground, requiring temperature compensation for any sensitive electronics.

If the DAMA/LIBRA measurements do indeed result from recoils, the data indicate that the WIMPs have a mass of about 100 protons and an interaction strength large enough so that other WIMP detectors should have seen nuclear recoil interactions. That other WIMP detectors have not observed recoils from dark matter remains a puzzle for the dark matter experimental community to solve.

A new generation of detectors just starting to take data may resolve the disagreement one way or the other. Several independent groups have built and operated a prototype detector to search for the DAMA/LIBRA signal. The most advanced, COSINE-100, is located in an underground laboratory in South Korea and has been taking measurements for 1.7 years. COSINE-100's background rates are low enough that the experiment should detect the DAMA/LIBRA effect in 5 years of operation.

The recoil direction also provides very important information about the solar system's motion through the dark matter halo. The direction of a nuclear recoil depends on the incoming direction of the dark matter particle that struck it. Viewed from Earth, the constellation Cygnus follows a circle in the sky, so the change in WIMP direction would occur daily, rather than yearly. A WIMP striking a stationary nucleus will cause the nucleus to recoil in the same direction that the WIMP was traveling. Measuring the recoil direction of the nucleus measures the WIMP direction.

However, measuring the WIMP direction is quite difficult: A recoiling nucleus will travel a few millionths of a meter in solid material and a fraction of a millimeter in a gas at atmospheric pressure. In a gas at one-twentieth of atmospheric pressure, a recoiling nucleus will travel a couple of millimeters, sufficient to make a direction measurement. The recoiling nucleus knocks electrons off the surrounding gas atoms as it recoils, leaving a pattern of its trajectory in free electrons. An electric field then pushes the free electrons through the gas, preserving the recoil pattern, to an amplification system that measures the direction the recoil nucleus followed.

Constructing a directional detector with a ton of target material and measuring more than 10 meters on a side presents a daunting, but feasible, task. If one of the current WIMP detectors observes a significant signal ascribable to dark matter, building a large directional detector will be essential. Observing the directional variation in the incoming WIMP direction would provide smoking-gun evidence for WIMP dark matter, beginning an era of dark matter astronomy.

WIMPs are the dominant paradigm for dark matter. The idea for these particles grows out of well-studied extensions of the Standard Model and provides an alternative explanation for the number of baryons produced in the Big Bang. The only real problem with WIMPs is that they have not been observed.

6
SEARCHING FOR DARK MATTER IN SPACE

Chapter 5 described how experiments on Earth search for WIMPs. This chapter explains how the annihilation of dark matter in our galaxy could produce cosmic rays—energetic normal matter particles traveling through space. Processes like supernova explosions and pulsars are known to produce cosmic rays, and astrophysicists have been observing cosmic rays for more than 100 years. Experiments searching for evidence for dark matter in cosmic rays must separate the cosmic rays resulting from dark matter annihilation from those produced by supernova explosions, pulsars, and other sources.

Section 6.1 describes how WIMPs could annihilate in the Milky Way to become Standard Model particles that then traverse the Milky Way. Section 6.2 deals with the detection of cosmic rays in space and gives the results from the most recent experiments. The chapter concludes with an unexpected observation resulting from the search for dark matter at the center of our galaxy.

6.1 WIMP ANNIHILATION IN THE GALAXY

Many theories of dark matter predict that dark matter particles will annihilate one another to produce Standard Model particles. The interaction rate on Earth for both WIMPs and axions in the galactic halo depends on the number density of the dark matter particles in the galactic halo. Going back to the parking lot analogy Section 5.2 in Chapter 5, imagine a lot with no parked cars and several blindfolded drivers driving around. There would be an average time for any two cars to collide. If we doubled the number of cars driven by blindfolded drivers, careening around the parking lot, each car would be twice as likely to suffer a collision, so the average time for a collision to occur would be four times shorter than before the number of cars was doubled. Think of the cars as being the dark matter particles moving around the galaxy in all different directions. Since the interaction rate is the reciprocal of the average time for any collision to occur, the collision rate between dark matter particles in the galaxy is proportional to the square of the number density of the particles at a given place in the galaxy.

The annihilation rate also depends on the probability of annihilation if the dark matter particles collide. Experimenters do not know the annihilation probability, but models of dark matter predict the annihilation probability. Experimenters do have some idea of the dark matter density and flux. The interaction probability between dark matter and normal matter, important for the Earthbound experiments described in Chapter 5, may be so low that an Earthbound experiment would not observe dark matter for a long time. The density-squared dependence of dark matter annihilation in the galaxy could make cosmic rays the best place to look for evidence of dark matter.

Most models predict that dark matter particles will annihilate with one another to produce energetic Standard Model particles that we can detect. Dark matter annihilation and the decay of the products from the annihilation would produce very energetic photons, protons, electrons, and neutrinos, and their anti-particles. In a typical model, two dark matter particles annihilate to produce two Z^0 bosons. First observed at CERN in 1983, Z^0 bosons are the weak force carriers in the Standard Model. The properties of Z^0 bosons are well known: They are unstable, and after a fraction of a yoctosecond (10^{-24}s), decay into pairs of electrons, muons, tau leptons, quarks, or neutrinos, and their anti-particles. The muons, tau leptons, and quarks subsequently decay into stable normal matter decay products: electrons, positrons, photons, protons, neutrinos, and their anti-particles. These decay products then travel across the galaxy. The dark matter annihilation and subsequent decay convert the dark matter particle mass into kinetic energy of the stable decay products, leading us to expect that the decay products will have high energies compared with most cosmic rays from known sources. Measurements of rotation curves described in Chapter 2 indicate that the dark matter density is higher at the center of galaxies, so we would expect more annihilations there. The energetic normal matter products from dark matter annihilation can travel across the galaxy. Some of these cosmic rays would travel close to Earth, where we hope to detect them. First, though, we need to see how the trip from their creation to Earth, which could be thousands of light-years, changes the cosmic rays' energy and composition.

When a dying star collapses and explodes into a supernova, the explosion sends a shock wave out and away from the star into the surrounding space. Space near a star contains a few protons per cubic centimeter. Enrico Fermi worked out how the expanding shock wave could accelerate protons, electrons, and atomic nuclei to very high energies. Supernovas happen all over the Milky Way, filling it with these energetic particles—cosmic rays—that cross the Milky Way; and some of these particles impinge on Earth.

Interstellar space is a complicated place: There is a weak magnetic field of a few μG, one-millionth of the Earth's magnetic field; and near Earth, there are about 30 protons per liter of "empty" space.[21] Stars and neutron stars produce larger magnetic fields throughout the galaxy, and the magnetic field extends far from a star. A charged particle like a proton or electron will be deflected by the galactic magnetic field, causing the particle to orbit around the field. Since the magnetic field does not affect the cosmic ray's velocity along the magnetic field, the particle must follow the field in a big spiral (Fig. 6.1). The magnetic field stretches from star to star, and the particles move along the field until they reach a star, where they whirl around in the big dipole magnetic field of the star and then leave again, headed for another star, still following the magnetic field. The particle's speed is very close to the speed of light, but part of its velocity goes into whirling around in the magnetic field, so that charged cosmic rays travel along the magnetic field at about half the speed of light. Stars in the vicinity of Earth lie a few light-years apart, so charged cosmic rays pass close to a star every few years.

In between encounters with stars, lots of things happen to the cosmic rays as they travel across "empty" space. Charged cosmic rays that pass close to a hydrogen atom will

---

21. A liter is about the same as a quart.

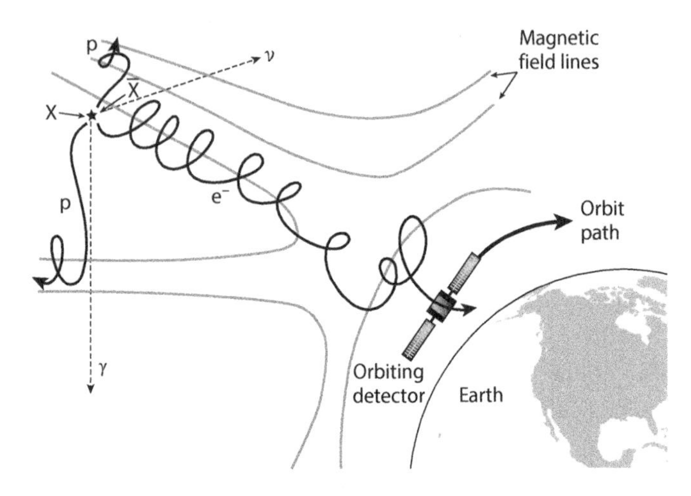

Figure 6.1. Detecting dark matter annihilation products in space. An energetic electron, produced in the annihilation of a dark matter particle x and anti-particle $\bar{x}$, is shown orbiting a magnetic field line and being guided by the line. The magnetic field in the galaxy measures a few μG, and the proton orbits with a radius of 10 million kilometers, eventually passing through the spaceborne detector shown. Neutrino and photon trajectories are not influenced by the magnetic field. They follow straight lines through the galaxy.

occasionally lose some energy by knocking the electron out of the atom. This process is called **ionization energy loss**. Cosmic ray protons also can hit a hydrogen nucleus directly and break it apart, creating a spray of lower-energy stable particles. Cosmic ray electrons, positrons, and photons also may crash into the hydrogen nucleus, but they create a mass of lower-energy electrons, positrons, and photons. All of these different ways in which cosmic rays interact with the interstellar hydrogen cause a "blurring" of the cosmic ray spectrum: Some of the protons in a single energy beam would hit protons and helium atoms floating in between stars and lose energy or, in rarer catastrophic collisions, would produce electrons, positrons, gamma rays, neutrinos, or other normal matter particles. Even the protons that emerge unscathed from a long trip across the galaxy would lose some energy from periodic ionization of the interstellar hydrogen and helium.

Aside from any possible products of dark matter annihilation, there are already many particles flying through space. About 3,000 protons from known sources like supernovas hit each square meter of the upper atmosphere every second, along with 300 electrons, 30 positrons, and 0.6 anti-protons.

In cosmic rays, one observes many more electrons than positrons—a ratio of about a thousand to one. The few positrons are created by interactions of the electrons during their acceleration by the shock wave. Theories of dark matter annihilation predict that the annihilations produce equal numbers of electrons and positrons, and equal numbers of protons and anti-protons. The annihilations also should produce energetic photons or gamma rays. Extra positrons, anti-protons, or gamma rays in cosmic rays could be an indication of dark matter annihilation taking place in the center of the Milky Way.

6.2 DETECTING COSMIC RAYS

Two things make cosmic rays difficult to measure on Earth: the Earth's atmosphere and its magnetic field. To reach a detector on the surface of the Earth, a cosmic ray must pass through 65 kilometers of the atmosphere, which is mostly nitrogen. Any cosmic ray will almost certainly hit an atom in the atmosphere and create other shower particles (see Fig. 5.5), severely diluting the signature for dark matter. Additionally, the Earth's magnetic field, which extends out a few thousand kilometers from Earth, bends cosmic rays away from the Earth. The Earth's magnetic field alters the cosmic ray's direction, making identifying evidence for dark matter annihilation more difficult.

These two obstacles mean that a cosmic ray detector must be either high in the atmosphere on a balloon or in orbit outside the Earth's atmosphere. Balloons have the advantages of being relatively simple and inexpensive to operate, and they can carry large payloads of up to several tons. However, the exposure time is limited to about 10 to 50 days by current balloon technology, and the balloons do not get completely outside the atmosphere. They get near the top of the atmosphere, but the remaining 1% of the atmosphere above the balloon still can cause interactions that mimic dark matter signals.

Spaceborne experiments present an alternative to balloon-borne experiments. Low Earth orbit (160 km to 2,000 km) is well above the atmosphere; and, once in orbit, an experiment can operate for years. These advantages can compensate for the severe restrictions on weight and electrical power imposed by space flight.

Cosmic ray experiments, whether on a balloon or in space, require a large magnet with a trajectory measuring system, called a **tracker**, inside the magnet to measure the sign of the electrical charge of the particle, which tells whether it is a particle or anti-particle; in addition to several other detectors above and below the magnet along the particle's trajectory to measure the particle's velocity and type (Fig. 6.2). The magnet presents the most significant experimental challenge: It must be large, something like a meter in length and width, to allow a large number of particles to pass through; and the magnetic field must be strong enough to bend the particles during their passage, to measure the momentum of each. If the magnet is superconducting, and the balloon payload or satellite also must carry enough liquid helium coolant to keep the magnet cold throughout its flight.

The tracker must measure the particle's trajectory accurately enough to give a good measurement of the particle's momentum. The detector design also must ensure that particles pass through the detector from top to bottom, rather than from bottom to top (Fig. 6.2). To do this, precision timing detection above and below the tracker must measure the travel time of a particle with an uncertainty of less than 0.1 billionth of a second. Measuring the transit time t over a distance d gives the velocity, $v = d/t$. Using a measurement of the momentum p from the bend in the magnetic field, we can find the mass of these particles from Einstein's relation $m = p\sqrt{c^2 - v^2}/(vc)$. The determination of curvature of the particle's trajectory gives the sign of the charge, since the direction of the deflection left or right depends on the sign of the particle's charge. A typical experiment may have a magnetic field of 1 tesla, 30,000 times larger than the Earth's field and a length of 1 m, and it must measure positions along the track with an accuracy of 10 μm. This will give an accuracy of 10% for the momentum. A high-energy particle moving near the velocity of light will cover the 1 m between the timing detectors in

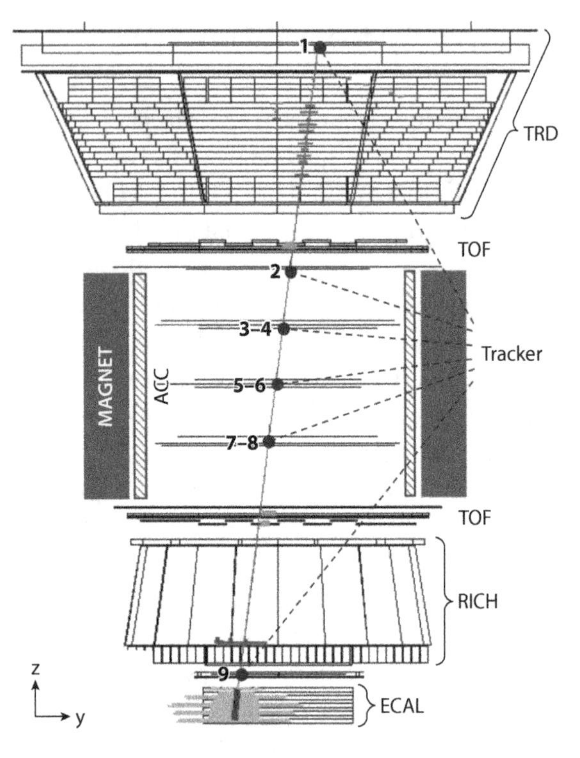

Figure 6.2. The charged particle magnetic spectrometer. Particles entering from the top first encounter a time-measuring detector called a **time of flight (TOF)** system. A particle's interaction with the TOF system starts a timer. The particle then passes through a **transition radiation detector (TRD)**, followed by a series of position-measuring detectors that determine the particle's momentum. A second part of the TOF system stops the timer when the particle passes through, and the time between the two TOF detectors gives the particle's velocity. The **electromagnetic calorimeter** and **ring imaging Cerenkov detector (RICH)**, along with the TRD, identify the type of particle and measure its energy.
Source: www.ams02.org. Reprinted figure with permission from as follows: M. Aguilar et al., "First Result from the Alpha Magnetic Spectrometer on the International Space Station: Precision Measurement of the Positron Fraction in Primary Cosmic Rays of 0.5–350 GeV," L 110, 141102 (2013) by the American Physical Society.

$1\ \text{m}/(3 \times 10^8 \text{m/s}) = 3$ ns, so a timing uncertainty of 100 ps translates to a velocity measurement of 3%. This system will measure particle masses to about 5% accuracy, sufficient to distinguish protons from positrons at lower energies.

In the 1990s, a hint of a signal for dark matter appeared in data from two balloon-borne cosmic ray experiments: the High Energy Anti-matter Telescope (HEAT) and the Advanced Thin Ionization Calorimeter (ATIC). The hint came from careful measurements of the ratio of high-energy positrons to high-energy electrons. More than expected positrons were measured compared with electrons, which would be expected from dark matter annihilation. Unfortunately, because balloon experiments do not stay aloft for very long, HEAT and ATIC did not collect enough data for conclusive results.

Several recent experiments have measured cosmic rays outside the atmosphere. The Payload for Anti-matter Matter Exploration and Light-nuclei Astrophysics (PAMELA) experiment was launched in 2006. By 2014, it had measured the positron/electron ratio, finding an increase at higher energies, in disagreement with earlier measurements. The additional positrons above prediction could indicate the annihilation of dark matter particles. PAMELA also measured anti-protons, finding approximately the number expected from supernova explosions. Taken together, the two measurements do not make sense for any of the several theories currently being considered, and new theories of dark matter or the way cosmic rays cross the galaxy are needed to explain these results—realizable theories of dark matter that predict both additional anti-protons and positrons in cosmic rays.

The Fermi Gamma-Ray Space Telescope, known as "Fermi," was launched in 2008 and measures both gamma ray photons and charged cosmic rays up to high energies with the Large Area Telescope (Fermi/LAT). Like ATIC, Fermi has no magnet, but the Fermi scientists found a clever way of using the Earth's magnetic field to separate electrons and positrons. The Fermi team used the fact that the Earth's magnetic field bends more positrons to the West than to the East to make a statistical measurement of the number of positrons in cosmic rays. After a year of data collection and analysis, the Fermi results covered the same region as PAMELA and ATIC but showed no evidence for an excess in the total electron-positron flux.

The Alpha Magnetic Spectrometer (AMS) was installed on the International Space Station in 2011 and has been collecting data ever since. AMS measurement of the positron/electron ratio agrees with PAMELA's measurement and extends it to higher energies. These results are consistent with theories of dark matter annihilation. However, the positrons also could be produced in the strong magnetic fields of pulsars. AMS also measures anti-protons and is finding more positrons at higher energies than predicted. Taken together, these results challenge both theories of dark matter annihilation and conventional models of particle production in the galaxy.

Dark matter annihilation also produces high-energy photons, or gamma rays, that travel to Earth along straight lines. Since magnetic fields do not affect a gamma ray's trajectory, experiments that measure the gamma ray's direction will be able to determine where the gamma ray came from. Theoretical models indicate that the centers of galaxies have

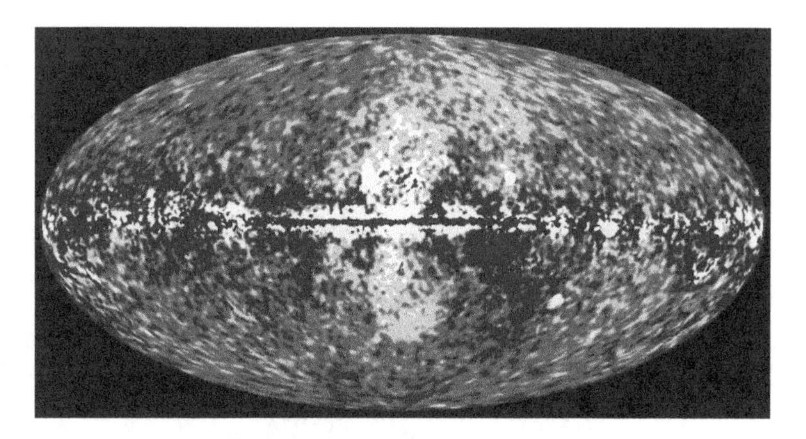

Figure 6.3. Gamma ray map of the galaxy from the Fermi/LAT. The disk of the galaxy lies along the center of the image. The grayscale in the map shows numbers of gamma ray photons from different parts of the sky above and below the galactic disk. The Fermi Bubbles are the oblong shapes extending about 25 klt-yr above and below the galactic disk.

high gamma ray densities, and observers look for photons from the center of the Milky Way or the centers of nearby dwarf galaxies. Supernovas, black holes, pulsars, and other astrophysical sources produce gamma rays not related to dark matter annihilation. Theorists and observers are working to develop a better understanding of these objects to improve the ability to separate gamma rays created by dark matter annihilation from those coming from other sources.

Recently, while measuring gamma rays from the center of the Milky Way, Fermi observers found two large gamma ray–emitting regions, or "bubbles," extending above and below the disk of the Milky Way (Fig. 6.3). Referred to as "Fermi Bubbles," since they were observed in data collected by the Fermi spacecraft, these structures extend 25 klt-yr above and below the plane of the galaxy. Based on careful measurements of the gamma ray energies, observers believe that the Fermi Bubbles formed in the last 10 million years, perhaps as the result of a large mass falling into the black hole at the center of the Milky Way and being ejected, or during a period of very active star formation near the center of the Milky Way. The Fermi Bubbles likely have nothing to do with dark matter, but they give a wonderful example of a serendipitous discovery.

The dark matter picture from cosmic rays remains quite complex and enigmatic: There is no unambiguous evidence for dark matter from any of the experiments. Most of the data does not fit with what theorists and modelers expect for cosmic rays from known sources or dark matter annihilation, stimulating a great deal of work in both improved modeling and dark matter annihilation theories. AMS and Fermi will continue to collect data. As a result, we will learn a lot about the Milky Way in the coming decade.

The cosmos may someday reveal the nature of dark matter to physicists and astrophysicists, but answers will not come easily. The early hopes of clear signals for dark matter from the energies of electrons, protons, or gamma rays have faded into the complexities of a zoo of occupants of the universe: black holes, pulsars, white dwarfs, and so on. However, astrophysical evidence for dark matter can give scientists clues of what to look for in terrestrial experiments.

7
SEARCHING FOR AXIONS

This chapter describes experiments to detect the axion, the subject of the next theory to explore. The idea originated in the 1980s, when theorists realized that the axion could be dark matter. Since then, many experiments have searched for axions with different masses and interaction strengths. The name "axion" comes from an Italian laundry detergent and was used by one of the particle's inventors, because the axion "cleans up" the strong interaction. Axions are very different from WIMPs: They are thought to interact mostly with photons, rather than nuclei; and their mass is at most a few millionths that of a proton.

Section 7.1 explains the theoretical ideas behind the axion; how axions interact; and how their mass, lifetime, and interaction strengths are related. Section 7.2 describes the most sensitive experiment searching for the lightest axion, and Section 7.3 describes a way of searching for the most massive axions by looking for their emission from the Sun. In recent years, there have been new experiments in both areas.

7.1 WHY DO WE NEED AXIONS?

The strong interaction's underlying theory, called **quantum chromodynamics (QCD)**, has an interesting feature: If $\mathcal{L}$

is a mathematical expression containing all the rules for performing calculations, then for the strong interaction, the theory may be written

$$\mathcal{L} = \mathcal{L}_{\mathrm{QCD}} + \theta \mathcal{L}_{\mathrm{CP}},$$

where $\mathcal{L}_{\mathrm{QCD}}$ represents the part of the theory that scientists have tested in accelerator experiments for 40 years; $\mathcal{L}_{\mathrm{CP}}$ represents another part of the theory that arises from its construction, and θ is a number between 0 and 2π. **CP** refers to reversing the direction time flows. Experimentally, QCD does not change if time is reversed.[22] Theory allows a term $\theta \mathcal{L}_{\mathrm{CP}}$ that changes when time is reversed. Experimentally, θ is small and could be zero: $\theta = 0$ means that QCD does not change if time is reversed.

$\mathcal{L}_{\mathrm{CP}}$ makes predictions that do not contradict any predictions of $\mathcal{L}_{\mathrm{QCD}}$ and so are allowed by the theory. For example, L_{CP} predicts that the neutrons line up in an electron field in a certain way from a property called the neutron electric dipole moment. However, experiments have failed to observe a neutron lining up in an electric field, and the results imply that θ is less than 10^{-9}.

For theorists, for a number that must be between 0 and 2π to be so close to or exactly zero is a hint that something is going on. In this case, a particle that is allowed but not predicted by $\mathcal{L}_{\mathrm{QCD}}$, called the axion, fixes the problem by continuously interacting with the Standard Model particles to ensure that $\theta = 0$ always.

---

22. One way of checking whether QCD changes when time is reversed is by comparing the rates of a reaction $a + b \rightarrow c + d$ with the reaction $c + d \rightarrow a + b$.

If the axion does exist, the Standard Model with axions allows the axion to decay to photons, $a \to \gamma + \gamma$, or for a photon to absorb an axion, increasing its energy by $m_a c^2$, where m_a is the axion mass, $\gamma + a \to \gamma$. The lifetime of an axion before it decays is inversely proportional to the axion's mass; and, if axions are dark matter, they must have survived from their creation in the Big Bang until now, limiting their mass to less than 20 billionths of a proton mass. Careful theoretical studies of the axion in the early universe indicate that its mass must be at least a quadrillionth—10^{-15}—of the mass of a proton for the axion to account for the effects attributed to dark matter. Limits on axion emission from red giant stars and super novas place an upper limit on the axion mass of about 10^{-12} of a proton mass Finally, the axion interaction strength is proportional to the axion's mass, implying that the axion has an interaction strength with normal matter for the mass range: 1×10^{-15} to 1×10^{-12} of a proton mass.

Experiments built to detect axions work in a very different way than WIMP detectors, because axions are a very different type of particle. If they exist, axions will have a tiny mass, which means that axion detectors need to provide a target of massless particles, photons, in the form of an electric or magnetic field. When an axion hits a target photon that is part of the target magnetic field, the axion converts to a photon that can be detected.

7.2 THE AXION DARK MATTER EXPERIMENT

The Axion Dark Matter Experiment (ADMX) uses an 8 tesla magnetic field as a photon target. According to quantum theory, any electric or magnetic field is composed of photons of different frequencies, and a static magnetic field is composed mostly of very-low-frequency photons. The target photon density depends on the square of the magnetic field strength, and a strong magnetic field will act as a dense photon target for axions that comprise the dark matter galactic halo. Using a to represent an axion, γ_M for the target photons in the magnetic field, and γ for a microwave photon, the reaction is $\gamma_M + a \rightarrow \gamma$. Since the target photons have so little energy, the frequency of the microwave photon depends on the axion mass. Because the experimenters do not know the axion mass, and so do not know the energy (or frequency) of the microwave photon their experiments are trying to detect, the experiments are designed to cover the range of predicted axion masses.

A cylindrical cavity will trap a microwave photon for several milliseconds if liquid helium keeps the walls of the cavity cold enough to superconduct, preventing the absorption of microwave energy. The superconducting walls of the cavity form nearly perfect mirrors that reflect the microwave photons into the cavity. The long microwave-photon trapping time will make electronic detection of the trapped photon possible. However, the electromagnetic theory states that the longer the trapping time for a cavity, the more precisely the microwave photon's frequency has to match the cavity's resonant frequency. In other words, an axion interaction makes the cavity ring like a bell, and the closer the frequency of the photon is to the bell's exact resonant frequency, the purer the tone of the bell. In the case of ADMX, for an axion with a mass 2×10^{-15} of a proton's mass, the microwave photon's frequency will be 460 MHz, and the resonant frequency of the cavity must be within 25 kHz, or 0.01% of the resonant frequency. Because the axion mass is unknown, ADMX must scan the frequency range from 460 MHz to 790 MHz, covering the axion mass range of $2–3.3 \times 10^{-15}$ of a proton mass. Since the photon produced in the absorption of the axion by the magnetic field must be within 25 kHz of the cavity's resonant frequency, ADMX must make 24,000 measurements in 50 kHz steps in the cavity frequency to cover the whole region where axions are thought to lie.

The very low energy of the microwave photons presents a second challenge. Collecting a detectable signal from an axion interaction requires accumulating trapped microwaves in the ADMX cavity over time. The microwave energy adds up over this "integration time" for each microwave frequency. During 1,000 s, several thousand axions might interact, putting 10^{-24} W into the cavity that ADMX can detect. The amount of power that ADMX is able to detect is about the same as that entering your eye from a 100 W light bulb 20 million kilometers away—ADMX is one of the most sensitive power detectors in the world.

Making 24,000 measurements when each one takes 1,000 seconds means operating the detector continuously for 24 million seconds, or more than 9 months. ADMX took even smaller steps, so the ADMX experiment will be operated for several years to scan the range from 300 MHz to 900 MHz.

Figure 7.1 shows a schematic of the ADMX experiment. An 8 T magnet, measuring 1 m in height, surrounds the microwave cavity, and a bath of liquid helium cools both

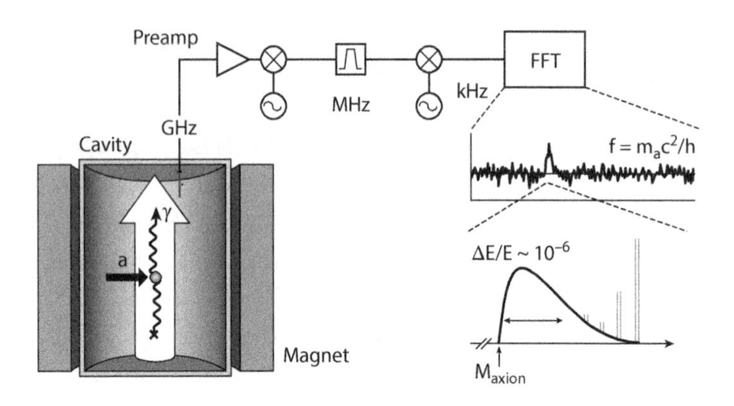

Figure 7.1. The ADMX experiment. The lower left of the figure shows the ADMX cavity, magnetic field, and an axion interacting with a photon in the magnetic field (indicated by the large, light arrow) and converting to a microwave photon trapped in the cavity. The cavity measures 1 m in height. A small antenna at the top of the cavity connects the cavity to the signal chain across the top of the figure that filters to eliminate noise. **Fast Fourier Transform (FFT)** is a means of decomposing an electronic signal into frequency components, shown below the signal chain. ADMX sums the frequency components over 1,000 s and stores the result. The red curve in the lower right shows the summed signal, assuming an axion with a mass at the cavity frequency.
Source: ADMX Collaboration. From "Status of the ADMX and ADMX-HF experiments," Karl van Bibberl and Gianpaolo Carosi (arXiv:1304.7803 [physics.ins-det], 2013).

the cavity and the magnet to superconducting temperatures to trap the microwave photons from axion interactions. An axion may pass through the cavity, interact in the magnetic field, and convert to a microwave photon. If the resonant frequency of the cavity matches the axion's mass, the cavity might trap the microwave photon. Movable dielectric rods adjust the cavity trapping frequency, making a frequency scan possible.

The rest of ADMX works somewhat like a fancy radio: An antenna in the cavity detects the microwave photon and sends the electrical signal to an amplifier. A tuning circuit selects the frequency, reduces noise, integrates the power picked up by the antenna over 1,000 s, and stores the measured power.

ADMX completed a scan of 460 MHz to 790 MHz in 2020 and did not detect an axion signal, excluding axions in the mass range of 2 to 3.3×10^{-15} of a proton mass from being dark matter. ADMX advanced lower-power microwave detection technology and showed that the method of photon trapping could detect axions at the rate predicted if axions were dark matter. Planned upgrades to ADMX that will extend the axion mass sensitivity to 1 to 40×10^{-15} of a proton mass will be operating soon.

In the past few years, experimenters have developed **broadband** axion detectors that use a large magnetic field as a target and are sensitive enough not to need a microwave cavity to capture the axions created in the magnetic field. Detectors with 10 cm magnets have demonstrated the broadband principle of scanning over 1 GHz, albeit with a sensitivity of about a fifth of the CERN Axion Solar Telescope's, described in the next section.

7.3 THE CERN AXION SOLAR TELESCOPE (CAST)

Stars like the Sun contain a large electric field that created the plasmas near the solar core. Similar to the magnetic field in ADMX, photons with nearly zero frequency compose the electric field. If axions exist, two photons from the solar electric field might fuse to form an axion: $\gamma_S + \gamma_S \rightarrow a$, where γ_S represents a solar photon. Axions created by this electric field photon fusion would have the energy of an x-ray and stream radially away from the Sun. X-ray photons have energies about a billion times higher than a microwave photon, so solar axion experiments can detect axions with a mass approximately a billion times larger than ADMX can.

Detecting solar axions works much the way ADMX detects lower-mass axions: An intense magnetic field provides a photon target, and the axion that has traveled from the Sun to the Earth hits a target photon, converting into an x-ray photon. The target photon comes from a static magnetic field, and the x-ray photon carries all the energy of the axion created in the Sun, following the same radial course away from the Sun.

A solar axion detector must detect the x-ray photon's energy and direction of travel, so that the experimenter knows that the x-ray resulted from an axion from the Sun and not some other x-ray source, of which there are many in the sky. Between 2003 and 2011, CAST applied its technology in the search for axions. CAST consisted of a 9.6 m dipole magnet from the LHC accelerator at CERN, with x-ray detectors mounted on one end of the magnet (Fig. 7.2). It rested on a mount that enabled the bore of the magnet to point at the Sun for about 90 minutes during sunrise and sunset. Tracking the Sun enabled CAST to make a differential measurement comparing the x-ray rate when the magnet bore pointed at the Sun with the rate when the magnet bore pointed near, but not at, the Sun.

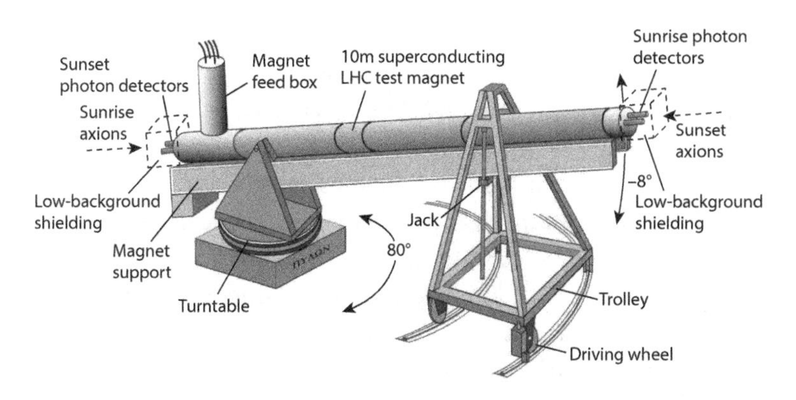

Figure 7.2. CAST. The long LHC test dipole magnet is aimed at the Sun. The dipole magnet has two bores, or channels, down the center. Axions enter from the left or right and convert in one of the bores of the magnet to an x-ray photon. The x-ray photon then travels to the end of the magnet, where it is detected by one of two x-ray detectors. The magnet could be adjusted by 8° in elevation and by 40° in azimuth, allowing the apparatus to be aligned such that one end of the magnet pointed at sunrise and the other at sunset. The magnet could track the Sun for 90 minutes at sunrise and sunset.
Source: From "Commissioning and First Operation of the Cryogenics for the CERN Axion Solar Telescope (CAST)" by Barth, K; Delikaris, D; et al (AIP Conf. Proc. 710 (2004) 168–175).

The advantage of making a differential measurement is that it is less sensitive to the background x-rays from radioactive contaminants in the detectors, and from cosmic ray interactions, since these backgrounds do not depend on whether the magnet is pointing at the Sun.

CAST observed the Sun in 2003 and 2004 for nearly 2,000 hours and did not observe an axion signal. Given the higher energy of the signal photons, CAST is sensitive to higher-mass axions than ADMX, and CAST would have detected heavier axions than ADMX (i.e., those with masses below 20 trillionths of a proton mass).

The second phase of CAST extended the mass range by adding ^{4}He gas to the bore of the magnet where the axions would hit the target photons and convert to x-rays. The helium effectively increased the mass of the target photons, making the conversion of higher-mass axions to x-rays possible and extending CAST's mass sensitivity to 400 trillionths of a proton mass. Adding helium to the conversion volume extended the mass range up to one-billionth of the mass of a proton.

CAST searched for axions from 2005 to 2007 using ^{4}He, and in 2009 to 2011 using ^{3}He, without observing axions in either run. The strength of the magnetic field and length of the magnet limited its sensitivity to values larger than those theoretically predicted for all but the heaviest axions. CERN currently plans a new axion search, called the International Axion Observatory (IAXO), for the coming decade. Experimenters are also using x-ray telescopes to look at nearby stars like Betelguese.

In the past few years, there have been new ideas for axion experiments that do not need to scan frequency, allowing longer collection times, which would increase power sensitivity. The greater power sensitivity means that the new experiments may be able to detect axions with smaller interaction strengths with magnetic field target photons.

Axions remain a compelling explanation for the dark matter problem and have important theoretical implications. I think that for a long time, axion experiments have taken a back seat to the nuclear recoil experiments that look for WIMPs, but compelling new ideas for experiments are attracting attention and resources. Following suggestions by theorists, experimenters have demonstrated broadband axion experiments that do not need the precise tuning that ADMX requires. Experimenters are now designing large-scale versions—a factor of 1,000 times larger and more sensitive—that, we hope, will produce exciting results in the next decade.

8
EPILOGUE

As you may have guessed, as the chapters have passed, physicists and astronomers have not yet learned what dark matter is. Still, as with the specters described in the Introduction, we have been able to identify some characteristics over time. We know that it is not any particle in the Standard Model; and nuclear recoil experiments have shown that a dark matter particle either weighs more than 1,000 protons or less than a fraction of a proton, with an interaction strength a billion times that of a neutrino. This chapter presents a summary of the dark matter experimental picture, as new theories have prompted experimental programs, described in the following section. We end with a consideration of where we might be in our quest to solve the dark matter problem in the future. As we consider all that has been accomplished so far, and look ahead, the future of dark matter research is exciting indeed.

8.1 LOOKING FORWARD: CURRENT AND UPCOMING DARK MATTER EXPERIMENTS

We can anticipate progress on many fronts, as enhanced resources become available for experiments. New nuclear recoil experiments like LZ or XENON10T, with target masses of several tons, will have sensitivities of greater than a factor of 10 over current experiments. The improved CDMS experiment will be able to detect WIMPs with masses that are a fraction of a proton's mass. The Large Hadron Collider (LHC) has had two successful data-taking runs, observing the Higgs boson in 2012. It searched for new particles of the same mass range that could be dark matter but did not find any candidate particles. For its new data-taking run, beginning in 2022, the LHC will be able to probe higher-mass particles with smaller interaction strengths. Its third run will use more intense particle beams in a focused effort to create and observe a particle that could be dark matter. If the LHC sees particles with properties consistent with dark matter in the coming years, then cosmic ray and nuclear recoil experiments will have a much better idea of the mass and interaction strength of potential dark matter particles. A new particle observed by the LHC would be a very big deal, whether the particle is dark matter or not.

The AMS-02 experiment will continue to collect cosmic ray data at the International Space Station at least until 2024. Analyzing and modeling AMS-02's cosmic ray data continues to be an enormous undertaking. The additional data may improve astrophysicists' understanding of cosmic rays well enough for AMS-02 to deliver a hint about the annihilation of heavy dark matter in our galaxy.

Measurements of gravitational lensing of stars in the LMC and Andromeda have excluded PBHs as dark matter unless they have masses of 10^{-11} to 10^{-14} that of the Sun—about the same mass range as the moons around

Jupiter. The brightening of background stars from gravitational lensing by PBHs in this mass range is only minutes or seconds, making observations difficult. Perhaps the new generation of ground and space telescopes will be able to carry them out, proving that PBHs cannot be the predominant form of dark matter.

Axions are not as well studied as WIMPs or PBHs. Candidate axion masses range from a million billionths to a few millionths of a proton mass, and only two experiments have reached the predicted axion interaction strength for narrow slices of the allowed mass range. New experiments like Dark Matter Radio (DMRadio) and IAXO will carry out broadband searches that may reach the predicted interaction strength. These results will come in a decade.

For WIMP, PBH, and axion searches, continued progress requires larger instruments with greater sensitivity and longer data collection times. As described in earlier chapters, detecting WIMPs and axions relies on providing "targets" in the form of atomic nuclei for WIMPs to hit, or magnetic field volumes to convert axions to electromagnetic radiation. For both, larger detectors will provide more targets. For PBHs, larger telescopes will allow more frequent imaging of more stars, the targets for the lensing transients. In all three cases, the ability to detect dark matter improves in proportion to the detector or telescope size.

Our ability to detect dark matter is also proportional to exposure or data collection time for WIMP, axion, and PBH searches. More data collection time increases the number of recoils for WIMPs, conversions for axions, or transients for PBHs. Usually, scientists run an experiment while planning and building the next, better version, or planning an upgrade to the running experiment.

The typical time for new dark matter experiments is 3 to 10 years, so the data collection time is typically 3 to 10 years. In other words, stay tuned!

While larger experiments and longer running times impact WIMP, axion, and PBH searches in the same way, sensitivity changes do not. For WIMP searches, improvements in sensitivity translate into lower energy thresholds, resulting in exponentially more events. A small improvement in a WIMP detector's sensitivity can double the likelihood of a WIMP detection. Axion experiments rely on detecting microwave or x-ray photons on top of background noise. Narrowing the bandwidth to detect a signal, or reducing the background noise, improves the sensitivity for detecting axion conversions. Both improvements will give proportionally better experiment performance.

PBH searches using the gravitational lensing of starlight by a transiting PBH will improve with the telescope's size, as the telescope collects more light from each star or measures the amount of starlight more precisely. In both cases, the telescope can detect a smaller increase in the star's brightness from the lensing during the transit. However, most PBH searches will benefit more from being able to sample more stars. With established technologies, the law of diminishing returns for improvements sets in after a few generations; and with dark matter experiments in their fourth generation, technical progress has become expensive.

8.2 OUTLOOK

Scientists have been searching for dark matter for almost 40 years. WIMP searches have shown dark matter in a mass range of 1 to 1,000 proton masses cannot exceed an interaction strength more than about a billionth of the neutrino's interaction strength. Microwave axion experiments have ruled out small regions of predicted masses and interaction strength, and they will probe more mass and strength regions as time passes. PBH searches have ruled out PBHs from the mass of a comet to 30 times the Sun's mass for PBHs of a single mass, but not for PBHs with a range of masses. Experiments will continue on all fronts, and dark matter searches remain top-priority experimental and theoretical work.

Where does the unfinished search for dark matter lie, compared with the durations of the other great quests of particle physics? Einstein predicted gravitational waves in 1916, the first evidence appeared in the early 1970s, and the Laser Interferometer Gravitational-Wave Observatory (LIGO) made the first observation of a gravitational wave in 2016—a century after Einstein's prediction. British theoretical physicist Paul A. M. Dirac laid the foundations for relativistic quantum field theory in 1927, Italian physicist Enrico Fermi created a model for weak interactions in 1932, and experimental evidence of the unification between the electromagnetic and weak forces came in 1978, 46 years later. Following electroweak unification, completing the Standard Model by observing the top quark in 1993 and the Higgs boson in 2012 took about 34 more years. In modern times, decades are not a long time to solve a significant problem like discovering the nature of dark matter.

Yet the dark matter problem is different: There is no reason dark matter particles have to interact with normal matter, except by gravity. Theorists have solid grounds, based on extensions to the Standard Model, to believe that dark matter does interact with normal matter in ways other than gravity; but these are ideas, not facts. The interaction strength between dark matter and normal matter could be small or zero (except for gravity). The work so far has only shown that dark matter interacts with normal matter with a strength bounded from above—an interaction strength less than a particular value—based on the non-observation of dark matter interacting with normal matter, aside from gravitationally. This **single-sided bound** means experiments have to continuously improve their sensitivity until dark matter is found; and each improvement means a larger experiment, longer data collection times, new technological advances, or all three.

Another process, neutrinoless double beta decay, $0\nu\beta\beta$, presents a similar challenge as dark matter. The decay of $0\nu\beta\beta$ only occurs if the neutrino is its anti-particle, called a Majorana neutrino, and has a nonzero mass, called the Majorana mass. The rate of $0\nu\beta\beta$ decay depends on the square of the Majorana mass. About 70 years of $0\nu\beta\beta$ experiments have not observed this decay and have bounded the Majorana neutrino's mass from above. Still, the Majorana mass could be tiny, or the Majorana neutrino may not exist at all.

Like dark matter searches, several generations of $0\nu\beta\beta$ decay experiments have passed without detection, and next-generation experiments of both dark matter searches and $0\nu\beta\beta$ decay have begun. Both dark matter searches and $0\nu\beta\beta$ decay experiments address fundamental questions: the composition of 27% of the mass of the universe for dark matter searches, and the nature of the neutrino for $0\nu\beta\beta$. The nature of dark matter and Majorana neutrinos' existence are not MacGuffins.[23] They are keys to physics progress. These single-bounded problems present two of the essential questions facing science.

Despite the immense difficulties in searching for the dark matter described in this book, and the large uncertainties outlined in this chapter, science always provides hope: the new idea. Science holds out the promise of a new idea to solve a big problem, and science—coupled with human inspiration and ingenuity—frequently delivers on that promise. In the nineteenth century, scientists struggled with increasing contradictions of Newton's mechanics before Einstein came up with relativity theory and Bohr, Dirac, and other quantum mechanics physicists delivered quantum theory to us. When we teach relativity and quantum mechanics to our students today, we make these ideas seem obvious and inevitable, but success came after decades of plodding and confusion in both cases. And that is where things stand with dark matter—we are painstakingly working through powerful new experiments that could detect dark matter for the first time, with the ever-present hope of a breakthrough that could create a new path to discovery.

Thirty years ago, my colleague Jean-Luc Vuilluemier and I were washing lead bricks in nitric acid for a dark matter experiment we were building. As we dipped each brick into

23. A MacGuffin is an insignificant object that serves to advance a story line, originally used by Alfred Hitchcock, who said in a 1935 interview: "It might be a Scottish name, taken from a story about two men on a train. One man says, 'What's that package up there in the baggage rack?' And the other answers, 'Oh, that's a MacGuffin.' The first one asks, 'What's a MacGuffin?' 'Well,' the other man says, 'it's an apparatus for trapping lions in the Scottish Highlands.' The first man says, 'But there are no lions in the Scottish Highlands,' and the other one answers, 'Well then, that's no MacGuffin!' So you see that a MacGuffin is actually nothing at all."

the acid bath, the brick would change from dull grey to a shiny array of little patches of different textures. Looking at them, Jean-Luc said, "What I love about physics is there is always something interesting happening." While the experiments described in this book all hold promise, the key to the next great discovery could be something new entirely. The aspects of dark matter that remain unknown may become the inspiration for that "something interesting" that may lead to an unexpected and exciting outcome.

GLOSSARY

alpha decay a radioactive emission of a helium nucleus from a larger nucleus. 86

alpha particle a helium nucleus consisting of two protons and two neutrons, emitted in radioactive decays. Sometimes called an alpha ray. 86

alpha ray a helium nucleus consisting of two protons and two neutrons, emitted in radioactive decays. Also called an alpha particle. 106

anti-matter distinct from matter: anti-neutrons, anti-protons, positrons, and anti-neutrinos are anti-matter particles. 83

astronomer a scientist who studies the universe by making observations using telescopes. 2

astronomical unit (AU) a unit of measure equal to the distance from the Earth to the Sun: 149,597,871 kilometers (or 92,955,807.3 miles, or 8.3 light minutes). The AU provides a convenient unit of measure for distances in the solar system: From the Earth to the Sun is 1 AU, Mars lies at 1.52 AU, to Saturn is 9.5 AU, and to Pluto (on average) is 39.5 AU. 12

astrophysicist a scientist who uses physics to understand objects in the cosmos. Astrophysicists also may make astronomical observations. 2

atmospheric neutrino a low-mass, neutral particle created in the interaction of a cosmic ray with a nucleus in the atmosphere. 115

axion a very light particle hypothesized in an extension to the nuclear interaction. The axion could account for dark matter. 4

baryon a class of particles containing three quarks that include protons and neutrons. About 575 types of baryons have been observed at accelerators and in cosmic rays, but only protons and neutrons form stable nuclei. While protons never decay, an unbound neutron will decay after about 15 minutes. 65

beta ray an electron or positron emitted in radioactive decay. 86

Big Bang nucleosynthesis the production of the lightest nucleons in the first few minutes of the Big Bang. 84

broadband a device or communication channel that works with signals across a wide range of frequencies. 141

Bullet Cluster the smaller of two colliding clusters of galaxies in 1E 0657-56. Some astronomers and others use the term to refer to both clusters. 46

caustic a region, usually linear, in a wave of particles or fluid of much higher density than the surroundings. A wave moving into a changing environment can develop a caustic. 101

Cepheid variable stars a class of stars whose brightness varies with a frequency that depends on the star's mass. 21

Coma Cluster a cluster of 1,000 galaxies 323 Mlt-yr from Earth, spanning about 4 degrees on the sky. 30

cosmic microwave background (CMB) radio waves that pervade the universe. The CMB was visible light that began freely propagating when hydrogen formed 370,000 years after the Big Bang, rendering the universe transparent to radiation. As the universe expanded, the CMB photons stretched from visible to microwave wavelengths. 52

cosmic ray an energetic particle—such as a proton, photon, or neutrino—traversing the Milky Way. 83

cosmological constant a numerical factor Einstein added to his original field equations to described a universe in steady state. A specific form of dark energy that has its pressure equal to the opposite of its density. 20

cosmologist a scientist who studies the universe as whole. Cosmologists usually also are astrophysicists. 12

CP charge-parity symmetry. In most theories, charge-parity symmetry means the theory would make the same predictions if time were flowing in the opposite direction. One way of thinking about "time flowing in the opposite direction" can be thought of as taking a movie of a process and seeing if the theory would make the same predictions if the movie were shown backward. 136

dark energy a constant energy density pervading the universe, causing the expansion of space-time. The origin of dark energy remains unknown. 6

dark matter matter not associated with the production of light in stars. Dark matter concentrates around galactic clusters and galaxies, along with normal matter. 2

density fluctuations regions of space in which the dark matter density is higher or lower than the average dark matter density because of quantum mechanical uncertainties. 52

deuteron a proton and a neutron bound together to form an isotope of hydrogen. 84

double beta decay a very rare process in which an atomic nucleus emits two electrons. There are two kinds of double beta decay: one in which two neutrinos are emitted along with the two electrons, and another type without neutrino emission. 106

dynamical law a law that describes how forces are produced. 6

electromagnetic calorimeter a detector that measures the energy of electrons and gamma rays. 130

energy density the amount of energy per unit volume at a certain location. 24

energy threshold when measuring a particle with a detector, the minimum energy the particle must have to be detected. For a particle collision, the minimum total energy in the center of mass the colliding particles must have to produce a given final state. 118

Fast Fourier Transform (FFT) an algorithm for converting a time-varying electrical signal into its frequency components. 140

fermion a class of matter particles with the characteristic feature that two identical such particles—electrons, for example—cannot be in the same place with the same momentum. 64

fine-tuning problem the requirement that Standard Model parameters must have precise values for the mass of the Higgs boson to have its measured value. 72

force application of a force to a body changes the body's momentum. 3

force carrier a particle, such as a photon, that communicates a force from one matter particle to another. 65

gamma ray a very energetic photon, typically emitted in nuclear transitions. 106

gauge boson a boson that carries a force between two particles. The term "gauge" refers to a specific class of particles, incorporated in the Standard Model, that result from a quantum mechanical degree of freedom. The photon, gluon, and weak force carriers are all gauge bosons. 66

generation in the Standard Model, similar particles occur in pairs, and each pair is a generation. For example, the lowest mass generation of quarks is the (up, down) generation. 64

gravitational lensing the distortion of an image caused by the light from the image passing through a gravitational field between the light and an observer. 40

gravitational well a region of large gravitational attraction caused by the presence of matter; also called potential well. 54

hadron a sub-atomic particle composed of quarks. For example, baryons have three quarks, and mesons have a quark and an anti-quark. 64

Higgs boson a particle in the Standard Model with about 125 proton masses that pervades all of space, interacting with every particle constantly, giving the particles their mass. 66

Hubble flow the local motion of galaxies expanding with the universe. 27

inertia the tendency of an object in motion to remain in motion and an object at rest to remain at rest. 6

inflation the rapid expansion of the universe from 10^{-36} to 10^{-32} s after the Big Bang. Inflation is caused by a hypothesized particle transitioning to its lowest energy state. 52

inflaton a hypothetical massive particle that causes inflation in the very early universe. 52

ionization energy loss the loss of energy of a high-speed, charged particle by pulling an electron off a nearby atom. 126

kinematic law the study of the fundamental properties of motion. A kinematic law relates motion to applied forces. 6

Large and Small Magellanic Clouds (LMC and SMC) two dwarf galaxies 163 lt-yr outside the Milky Way galaxy. Gravity holds them to the Milky Way, making the LMC and the SMC satellites of the Milky Way. 13

light-year the distance traveled by light in one year (abbreviated lt-yr). One year is about 3×10^7 seconds. The speed of light is 3×10^8 m/s; so in one year, a light ray will travel about 9×10^{15} m. 12

mass-energy used to signify the stuff of the universe. Mass-energy includes the rest mass and kinetic energy of the "stuff." 12

Massive Compact Halo Objects (MACHOs) an astrophysical object whose size is small enough and whose density is high enough to cause microlensing of light from a star. PBHs are MACHOs. 89

matter Two uses: First, the matter particles in the Standard Model—quarks and leptons. Quarks and leptons bind in various ways to form matter. Second, distinct from anti-matter: Neutrons, protons, electrons, and neutrinos are matter particles. The first use includes anti-matter. 2

matter density the amount of matter per unit volume at a specific location. 25

microlensing the apparent brightening of a star or other luminous object by an intervening object of planetary to stellar mass. 46

modified Newtonian dynamics (MOND) a theory proposing a gravitational force law different from Newton's force law in an attempt to explain the effects of dark matter. 46

muon veto a particle detector that identifies muons and turns off another detector to prevent backgrounds caused by cosmic ray muons. 108

nebula from the German word for "foggy," nebula referred to extended astrophysical objects later called galaxies. 21

neutrino a chargeless, very light particle that interacts only weakly and gravitationally. There are three different kinds of neutrinos, each associated with a charged lepton. 65

normal or luminous matter matter bound in stars or dust clouds that emits or reflects light. Also referred to as visible matter. 2

particle annihilation the collision and conversion of a particle and its anti-particle to another form of matter. Electrons and positrons may collide and annihilate to form photons, neutrinos, or other particles, depending on the energy of the collision. 79

particle physicist a scientist who either develops theories or performs experiments to understand the interactions between fundamental particles. 2

particle theory the branch of particle physics tasked with developing formal theories that comprehensively explain experimental observations. 28

phase transition a change in state of matter from one form to another. Water freezing into ice is an example of a phase transition. 52

phonon a propagating vibration in a lattice of atoms. Phonons can be described quantum mechanically as particles. 109

primordial black holes (PBHs) a black hole formed in the early universe by quantum mechanical density fluctuations. 3

quantum chromodynamics (QCD) also known as the strong interaction, QCD is the force that binds the nucleus together. 135

quark a point-like particle found in protons and neutrons. Quarks are fermions. 3

quasar short for quasi-stellar radio source; the most distant observable objects, probably super-massive black holes in the centers of large galaxies. 44

recoil particle a stationary particle struck by an energetic particle, receiving momentum, and moving or recoiling. 101

redshift the stretching of light waves to longer wavelengths, toward the red end of the visible spectrum. 6

ring imaging Cerenkov detector (RICH) a charged particle emits a ring of Cerenkov radiation passing through low-density, transparent solids. A RICH collects the radiation and uses the measurement of the ring's radius to find the particle's velocity. 130

scalar boson a boson that does not carry angular momentum. The Higgs boson is a scalar boson. 66

scattering the collision and recoil of two or more particles. 54

scientific notation a means of expressing large numbers by indicating the number of zeros (the exponent) following a mantissa that gives value. Examples: $10^0 = 1$, $10^1 = 10$, $3 \times 10^3 = 3{,}000$, $6.5 \times 10^6 = 6{,}500{,}000$. 12

scintillation light light produced by charged particles moving inside certain materials, called scintillators, whose chemical composition make them produce optical photons from the ionization caused by a charged particle moving through the material. 112

single-sided bound an experimental or theoretical result that bounds a quantity from above or below. For example, accelerator experiments that unsuccessfully searched for the Higgs boson in the 1990s were able to place lower limits on the Higgs boson's mass. 151

spiral galaxy a galaxy with characteristic spiral arms. There are also elliptical and spherical galaxies. 13

Standard Model the reigning theory of fundamental interactions between particles. 3

supersymmetry (SUSY) a speculative theory that aims to solve the fine-tuning problem by postulating an as-yet-undiscovered partner particle for every particle in the Standard Model. 73

tau a cousin of the electron, with a mass 3,400 times larger. The tau lepton is unstable and can decay to a muon, electron, or hadrons, and a tau neutrino. 65

tidal force the stretching force felt by a body in a spatially changing gravitational field. The ocean tides on Earth partially result from the water on the side of the Earth closer to the Sun feeling a greater gravitational pull than the water on the side of the Earth farther away from the Sun. 10

time of flight (TOF) the time it takes for an energetic particle to travel between two detectors. A TOF measurement system has two detectors connected to a clock that measures the time a particle crosses each detector. 130

tracker a particle detector that determines the trajectory of a charged particle passing through it. When used in a magnetic field, it can determine the momentum of the particle. 128

transition radiation detector (TRD) a detector that measures the light emitted by an energetic charged particle when the particle crosses from one material to another. The measurement of the light gives a measurement of the particle's velocity. 130

vacuum expectation value (VEV) the quantum mechanics probability of finding a particle in the vacuum, the state where no other particles are present. In the Standard Model, the VEV for an electron is zero; while for the Higgs boson, the VEV is not zero. The nonzero VEV for the Higgs boson is part of the mechanism that causes all particles to have mass. 69

velocity dispersion for a collection of moving objects—galaxies gravitationally bound in a cluster, for example—the velocity dispersion is a statistical quantity that characterizes the spread in velocities around an average value. 31

virial theorem a theorem from mechanics that relates a mechanical system's kinetic and potential energies. 31

visible matter matter associated with the production of light in stars. 2

weak lensing gravitational lensing of a distant source by an intervening mass that produces a distortion of the image of the distant source but no multiple images. 46

Weakly Interacting Massive Particles (WIMPs) a particle not found in the Standard Model that could be dark matter; as yet unobserved. 4

white dwarf a collapsed star that has burned out all its fuel but does not have enough mass to be a neutron star or black hole. 46

SUGGESTED READINGS

Online free magazines provide the best places to keep up with dark matter happenings. These have extensive archives:

Physics (from the American Physical Society), physics.aps.org
Quanta, quantamagazone.org
Symmetry, symmetrymagazine.org

In 2003, *Science* devoted an entire issue to dark matter and energy:

Rowan, Linda, and Robert Coontz. "Welcome to the Dark Side: Delighted to See You." *Science* 300, no. 5627 (2003): 1893.

Some classics:

Sean Carroll, *From Eternity to Here: The Quest for the Ultimate Theory of Time* (Dutton, 2010)
Alan Guth, *The Inflationary Universe: The Quest for a New Theory of Cosmic Origins* (Perseus Books, 1997)
Richard Preston, *First Light: The Search for the Edge of the Universe* (Atlantic Monthly Press, 1987)
Steven Weinberg, *The First Three Minutes: A Modern View of the Origin of the Universe*, 2nd ed. (Basic Books, 1993)

INDEX

$E = mc^2$, 11, 78
Z^0 bosons, 124
1E 0657-56, 46–49, 51, 74, 96

acceleration, gravitational, 6, 8, 9, 11
  Moon, 11
Advanced Thin Ionization Calorimeter (ATIC), 131
alpha decays, 86, 102
Alpha Magnetic Spectrometer (AMS), 132
alpha rays, 87, *see also* beta rays; gamma rays, 106, 109
Alpher, Ralph, 76
AMS-02 experiment, 147
Andromeda galaxy, 13, 14, 21, 33, 34, 95, 147
  mass of, 38
  rotation of, 35
Andromeda Gravitational Amplification Pixel Experiment (AGAPE), 91
angular momentum
  dark matter, 99
annihilation
  anti-neutrons, 82
  anti-quarks, 80
  dark matter, 122–124, 126–128, 131–132, 134
  electron-positron, 81
  galaxy, 122–127
  neutrinos, 82
  neutron–anti-neutrons, 83
  probability for, 123
  rate, 123
anti-hadrons, 81
anti-matter, 83
anti-neutrinos, 80, 81, 82, 88
anti-neutrons, 81, 82
anti-particles, 81–83
anti-protons, 81–83
anti-quarks, 67, *see also* quarks, 80
  annihilation, 80
argon, 112
asteroids, 95
astronomers, 2, 3, 8, 12, 18, 21, 34, 40, 51, 61, 62, 85, 98, 99, 146
astronomical unit (AU), 12
astrophysicists, 2, 12, 122
atmospheric neutrinos, 115
atomic electrons, 36, 109, 110
atomic hydrogen, 52

atomic nuclei, 3, 125, 148
atoms, 63, 64, 105, 112, 115, 126
Axion Dark Matter Experiment (ADMX), 137–145
axions, 4, 73, 93, 94, 97, 98, 101, 123, 135–145, 148, 149
  detection, 135, 137
  interaction, 135
  magnetic field, 140
  mass, 138
  microwave, 150
  microwave photons, 140
  photons and, 135

Baade, Walter, 36
Babcock, Horace, 35, 36
baryonic compact objects, 88–92
baryonic gas, 99
baryonic matter, 85, 97, 99, 103
baryons, 14, 65, 82, 86, 88, 89, 92–94, 99, 100, 101, 116, 121
Becquerel, Henri, 4, 86
beta decay
  neutrons, 69
beta decays, 86, 87, 102
beta electrons, 87
beta rays, 86, 87, 106, 109
beta spectrum, 87
Betelguese, 144
BICEP2 experiment, 77
Big Bang, 28, 76
  light from, 51–60
  nucleosynthesis, 84
  protons in, 77
Big Bang theory, 28, 63, 76, 77, 85, 88, 97
black holes, 3, 46, 75, 76, 88, 92, 99, 133, 134
  formation, 76
  neutron star, 76
  primordial, 92–96
  supernova and, 76
blueshift, 18, 19
Bohr, Niels, 87, 152
Bohr's conjecture, 87
BOOMERanG experiment, 60
bosons, 64, 67, 68, 71–73
  exchange, 72
  gauge, 66
  scalar, 64, 66
bubbles, 133
Bullet Cluster, 46, 47, 49, *see also* Coma Cluster, 51

Carnegie Observatories, 36
caustics, 101
Cepheid variable stars, 21
CERN Axion Solar Telescope (CAST), 141–145
charged particle magnetic spectrometer, 130
clusters, 44, 61, 88
  motion of galaxies in, 32
clusters of galaxies, 2, 3, 9, 13, 20, 28, 29
cold dark matter, 61
Coma Cluster, 30–32, 38, 75
  mass-to-light ratio for, 30
  sky map, 31
  Zwicky's survey, 31

conservation of energy, 87
contraction-expansion-contraction cycle, 55, 57
COSINE-100, 120
Cosmic Background Explorer (COBE), 51, 58, 60
cosmic microwave background (CMB), 52, 58, 60, 76, 77, 85
cosmic rays, 83, 102, 105, 106, 107, 108, 111, 113, 114, 115, 122, 123, 124, 125, 147
 detection, 127–134
 Earth, 127
 electrons, 126
 experiments, 128
 muons, 108
 particles, 106
 proton, 114
 protons, 126
cosmological constant, 20, 27, 28
cosmologists, 12, 23, 77
cosmology, 28
cosmos, 6, 77, 134
Coulomb interaction, 54
Cowan, Clyde, 75
CP, 136
Cryogenic Dark Matter Search (CDMS), 110–112, 147
Cryogenic Rare Event Search with Superconducting Thermometers (CRESST), 111
cryostat, 108
Cygnus, 116, 117, 120
dark baryons, 89
dark energy, 20–28
 density, 26
dark matter, 20, 75, 105
 angular momentum, 99
 annihilation, 122–124, 127, 131, 132
 axions, 98, 141
 cosmic rays, 134
 defined, 2
 density fluctuations, 53
 detector, 108
 elastic scattering, 104
 experiments, 102, 116
 galactic halo, 123
 galaxies, 99–101
 gravitational field, 40
 gravitational forces, 29
 halo, 116–121
 interaction, 113
 interaction rate with normal matter, 103
 invisibility, 2
 neutrinos as, 86–88
 particles, 103
 PBH, 94
 potential well, 54, 55
 WIMPs, 102
Dark Matter Radio (DMRadio), 148
Dark Matter/Large sodium Iodide Bulk for RAre processes (DAMA/LIBRA) experiment, 118, 119, 120
decoupling, 82
density fluctuations, 52, 53, 57, 88, 98, 99

detector
cosmic rays, 128
nuclear recoil, 115
solar axions, 142
spaceborne, 126
deuterium, 53, 84, 88
deuterons, 84, 115
differential microwave radiometers (DMRs), 58
dipole magnet, 142, 143
Dirac, Paul A. M., 150, 152
distances in universe, 12–14
Andromeda galaxy, 13
astronomical unit (AU), 12
Earth to Moon, 12
Earth to Proxima Centauri, 13
Earth to Sun, 12
light-year (lt-yr), 12
meters, 12
scientific notation, 12
solar system to galactic superclusters, 15
Doppler shift of sound, 19
double beta decays, 106
dwarf galaxies, 34, 133
dying star, explosion, 124
dynamical law, 6

Earth
atmosphere, 115
atmosphere, 127
axions and, 123
cosmic rays, 127
magnetic field, 125, 128
Moon, distance to, 12
motion detection through dark matter halo, 116–121
orbit, 13, 99
size of, 13
Sun orbit, 13, 116
WIMPs and, 123
Eddington, Arthur, 43
Edelweiss-III, 108
Einstein, Albert, 4, 5, 11, 18, 23, 27, 43, 77, 150, 152
Einstein radius, 90, 91
Einstein's theory of general relativity, 23, 28, 77
Einstein's theory of gravity, 43, 72, 74, 152
electric field, 120, 138
electromagnetic calorimeter (ECAL), 130
electromagnetic decay, 69
electromagnetic energy, 68
electromagnetic forces, 3, 10, 29, 49, 72, 150
electromagnetic waves, 16, 44
electron-positron annihilation, 81, 83
electron-positron flux, 132
electronic forces, 68
electrons, 53, 65, 68, 80, 81, 86, 124
electroweak unification, 150
Ellis, Charles, 87
energy, 5–12
energy density, 23
energy threshold, 118
Expérience pour la Recherche d'Objets Sombres (EROS), 91, 92

failed stars, 32, 86, 88–89
Fast Fourier Transform (FFT), 140
Fermi, Enrico, 124, 150
Fermi Bubbles, 133–134
Fermi Gamma-Ray Space Telescope (Fermi), 132, 133
Fermi National Laboratory, 72
Fermi spacecraft, 134
fermions, 64, 67, 68, 71–73
fine-tuning problem, 72, 73
fission reactor, 75
forces, 3
  acting on body, 7
  bodies produces, 5
  carriers, 65
  defined, 6
  distance between the bodies and, 7
  electromagnetic, 3, 10, 29, 65, 72
  gravitational, 3, 5, 8, 15, 29
  motion of objects, 5
  size of, 7
  strong, 3, 29, 65, 72
  tidal, 10
  weak, 3, 29, 65, 72
Ford, Kent, 36, 38, 39
Friedmann, Alexander, 77

galactic halo, 123
galactic magnetic field, 125
galactic structure, 14
galactic superclusters, 15
galaxies, 2, 3, 21
  1E 0657-56, 46, 47–49
  Bullet Cluster, 46, 51
  dark matter in, 99–101
  formation, 38, 100
  lensing, 45
  mass of, 34
  motion of, 18, 30, 41
  motion of stars, 33
  New General Catalog (NGC), 39
  orbit of stars in, 32–38
  recession measurement, 22
  universality of rotation curves, 39
galaxy clusters, 13, 27, 61
Galilei, Galileo, 5
gamma rays, 106, 107, 108, 109, 111, 115, 127, 133, 134
  map of galaxy, 133
  photons, 110, 112, 132
Gamow, George, 76
gas, 78
  baryonic, 99
  thermal energy, 79
gauge bosons, 64, 66
generation of particles, 64
germanium, 104, 107, 109
  detector, 105
  double beta decays, 111
  experiments, 109, 110
gluons, 3, 65, 69, 78, 79
gravitational field, 40
gravitational forces, 3, 5, 8, 15, 29
gravitational lensing, 40–46, 50, 89, 147, 149
gravitational waves, 150
gravitational well, 54
gravity, 3, 4, 5, 67, 93
  bending of space and time, 72

gravity (*cont.*)
  dark matter and, 99
  distant bodies, between, 6
  effects, 5
  light ray deflection, 42
Green Bank Observatory, 36

hadrons, 64, 65, 67, 81
Hahn, Otto, 87
Hawking, Stephen, 92, 95
helium, 53, 127
Herman, Robert, 76
Higgs boson, 64, 66–72, 147, 150
  particles, 66, 69, 71
High Energy Anti-matter Telescope (HEAT), 131
high-energy electrons, 131
high-energy particles, 83
high-energy positrons, 131
HII regions, 36
Hitchcock, Alfred, 152
Hubble, Edwin, 19, 21, 22
Hubble Constant, 22
Hubble flow, 27
Hubble Space Telescope (HST), 16
Hubble's law, 22, 28
hydrogen, 54, 56, 57, 76
hydrogen atoms, 51, 65, 125
hydrogen gas, 34
  velocity of, 36

inertia, 6
inflationary theory, 77, 78
inflations, 52
inflatons, 52
International Axion Observatory (IAXO), 144, 148
intracluster gas, 47, 49
invisible matter, 3
ionization energy loss, 126
isotopes, 67

Kaptyn's star, 30, 32
Kepler, Johannes, 5
kinematic law, 6
kinetic energy, 15, 79, 87
  dark matter, 118
  nucleus, 102

Lambda-CDM model, 61
Large Area Telescope (Fermi/LAT), 132
Large Hadron Collider (LHC), 147
Large Magellanic Cloud (LMC), 13, 90, 147
Laser Interferometer Gravitational-Wave Observatory (LIGO), 150
LeMaître, Georges, 77
lensing effect, 43
leptons, 3, 4, 64, 65, 66, 68, 78–80, 124
Lick Observatory, 36
light detectors, 112
light waves, 17, 25
light-year (lt-yr), 12
liquid helium, 138, 139
liquid helium coolant, 129
Lowell Observatory, 18
LZ, 147

M31, *see* Andromeda galaxy
MacGuffins, 152
MACHO experiment, 89, 91, 92
magnet, 129
magnetic field, 125, 126, 138
Majorana mass, 151
Majorana neutrinos, 151
mass, 5–12
  axions, 137, 138
  measurement, 6
  protons, 137, 139, 141, 144
mass-energy, 12
mass-to-light ratios, 30
massive compact halo object (MACHOs), 89
matter, 2
matter density, 23
matter-energy density, 25
MAXIMA balloon experiment, 61
Mayall, Nichloas, 36
Meitner, Lise, 87
Mercury, 95
microlensing, 46
microwave axions, 150
microwave photon, 100
Milgrom, Mordechai, 96
Milky Way galaxy, 13, 21, 30, 89, 99–101
  lensing transiting objects, 90
missing mass, *see* dark matter
modified Newtonian dynamics (MOND), 46, 47, 74, 96
motion of objects, 5
Mount Palomar Observatory, 30
Mount Wilson Observatory, 36
muon veto, 108
muons, 64, 65, 66, 71, 114, 124
  spin frequency, 71

nebula, 21
neon, 112
Neptune, 13
neutral supersymmetric particles, 103
neutrinoless double beta decay, 151
neutrinos, 64, 65, 66, 75, 78–81, 86–88, 103, 111, 114–116, 124
  annihilation, 82
  atmospheric, 115
  dark matter, 86–88
  electron, 64, 87
  electron-positron annihilation to, 81
  muon, 64, 88
  tau, 64, 88
neutron stars, 75, 76, 89
neutrons, 3, 53, 65, 67, 81–83, 86, 87, 115, 136
  absorber, 108, 113
  beta decay, 69
  electric dipole moment, 136
New General Catalog (NGC), 39
Newton, Isaac, 5
Newton's dynamic law of gravitation, 7–8
Newton's law of gravitation, 5
  on distant object, 6
Newton's laws of motion, 5, 6
  first law, 6
  second law, 6, 8, 96
  third law, 7

noble gases, 112
normal matter, 2
nuclear beta decay, 75
nuclear physicist, 86
nuclear recoil experiments, 88, 144, 147
nucleus, 65
  recoiling, 110

Optical Gravitational Lensing Experiment (OGLE), 91, 92
orbital velocity, 34, 37, 38
orbits of stars in galaxies, 32–38
Ostriker, Jerimiah, 41

particle annihilation, 79
particle detectors, 105
particle physicist, 2
particle-anti-particle pairs, 83
particles decoupling, 82
Pauli, Wolfgang, 87
Payload for Anti-matter Matter Exploration and Light-nuclei Astrophysics (PAMELA), 131, 132
PBH, *see* primordial black holes (PBH)
Peebles, James, 41
phase transition, 52, 53
phonons, 109, 111
photo detectors, 113
photons, 16, 25, 40, 54, 59, 65, 68, 78–81, 109, 110, 114, 126, 132
  electron-positron annihilation, 81
  fusion, 80
  massless, 80
  trapping, 141
physicists, 75, 87
pion, 69, 100
Planck, 59, 60
planetary orbit, 10
planets, 88, 89
Pluto, orbit, 13
position, 8
position-measuring detectors, 130
positrons, 53, 80, 81, 83, 126
primordial black holes (PBH), 3, 92–96, 148, 150
primordial universe, 92
Principia (Newton), 6, 7
protons, 3, 53, 65, 67, 81–83, 86, 100, 114, 127
  cosmic rays, 114
Proxima Centauri, 13
pulsars, 122, 133

quantum chromodynamics (QCD), 135, 136
quantum field theory, 3
quantum mechanical density fluctuations, 88, 92, 99
quantum mechanical Standard Model for quarks and leptons, 4
quantum mechanics, 52, 87, 93
quarks, 3, 4, 64–68, 78–81, *see also* anti-quarks, 124
quasar, 44, 45

radio frequencies, 76
radioactive contamination, 106
radioactive decay, 4, 106
radioactivity, 107
radon gas, 107
recoil particle, 101, 102
redshift, 6, 18, 19, 76
  measuring speed using, 16–20
  orbital velocity, 34
  radio waves, 34
  velocities of galaxies, 22
Reines, Frederick, 75
relativistic quantum field theory, 150
ring imaging Cerenkov detector (RICH), 130
Roberts, Mort, 36, 38
Robertson, Howard, 77
Rubin, Vera, 36, 38, 39
Rutherford, Ernest, 87

scalar boson, 64, 66
scattering, 54
scattering off charged particles, 82
Schmidt-Cassegrain refracting telescope, 30
scientific notion, 12
scintillation light, 112
shock waves, 124
single-sided bound, 151
Slipher, Vesto, 18, 21, 33
Sloan Digital Sky Survey, 44
Small Magellanic Cloud (SMC), 13
solar axion detector, 142
solar system, 9, 116
solid-state detectors, 112
spaceborne experiments, 128
special relativity, 11
spiral galaxy, 13
spiral nebula, 33
spring scale, 10, 11
standard candles, 21
Standard Model of particle physics, 3, 4, 63, 64, 75, 79, 98
  axions, 137
  bosons, 67
  electrons, 66
  fermions, 67
  force carriers, 64–66
  forces, 65
  gluons, 78
  Higgs boson, 66–71
  leptons, 78
  neutrinos, 66, 78
  particles and interactions, 63–68
  photons, 78
  quarks, 66, 78
  testing, 71–74
Standard Model of particles physics, 123
stars
  failed, 88, 89
  light-emitting, 32, 33
  motion in galaxy, 33
  neutron, 75, 76, 89
  red giant, 137
  visible, 3
stellar nucleosynthesis, 84
strong forces, 3, 29, 72

Subaru telescope, 95
Sun, 13
super clusters, 13
superclusters, 13–15, 88
superconducting thermometers, 112
supernova, 76
supernova explosions, 122, 124
Supernovas, 133
supersymmetric particles, 97
Supersymmetry (SUSY), 73
SUSY, *see* Supersymmetry (SUSY)

tau, 64, 65, 124
theory of gravity, 4, 7, 27, 72, 94
Theory of Relativity, 18
  General, 5, 22
  Special, 5, 11, 18, 20, 22
thermal energy, 54
  gas, 79
thermal equilibrium, 80–82
thorium, 107
tidal force, 10
time of flight (TOF) system, 130
tracker, 128
transient lensing, 95
transition radiation detector (TRD) system, 130
Triangulum galaxy, 33, 34, 36
  rotation curve, 37
two energy measurements, 109, 110

universal laws of motion, 5
universe
  expansion of, 19, 22, 23, 26
  matter-energy constant, 22
Uppsala General Catalogue of Galaxies (UGC), 39
uranium, 107

vacuum expectation value (VEV), 69–71
vector bosons, 64, 68
velocity, 6, 8, 18
  dispersion, 31
  fractional recession, 26
  hydrogen gas, 36
  of galaxy, 21
  orbit, 34
  rotation of stars, 19
  stars in galaxies, 36
virial theorem, 31, 32
visible matter, 2
  gravitational forces on, 29
  motion of galaxies, 3
Volders, Louise, 36
Vuilluemier, Jean-Luc, 152

Walker, Arthur, 77
wavelength of light, 17
weak forces, 3, 29, 72, 150
weak lensing, 46, 47
Weakly Interacting Massive Particles (WIMPs), 4, 73, 93, 101, 102, 103, 104, 121, 144, 147, 148
  annihilation in galaxy, 122–127
  detectors, 107, 112, 115, 118, 119, 137

weight, 5–12
  Earth, on, 10
  force of gravity, 11
  Moon, on, 11
white dwarfs, 46, 88
Wilkinson Microwave Anisotropy Probe (WMAP), 59, 60
WIMP, *see* Weakly Interacting Massive Particles (WIMPs)
x-ray photon, 100, 102, 142, 143, 149
x-ray telescopes, 47, 144
xenon, 112
XENON10T, 147

Zwicky, Fritz, 30–32, 75
  galaxies of Coma Cluster, survey, 31
  measurements, 62